Recent Advances in Robot Manipulators

Edited by Serdar Küçük

Published in London, United Kingdom

Recent Advances in Robot Manipulators
http://dx.doi.org/10.5772/intechopen.100669
Edited by Serdar Küçük

Contributors
Mahonri William Owen, Chikit Au, Mihaiela Iliescu, Ciprian Dragne, Isabela Todirite, Marius Pandelea, Corina Radu Frenț, Petru-Alexandru Cotfas, Veturia Chiroiu, Dang Xuan Ba, Serdar Küçük, Burak Inner

First published in London, United Kingdom, 2022 by IntechOpen
IntechOpen is the global imprint of INTECHOPEN LIMITED, registered in England and Wales, registration number: 11086078, 5 Princes Gate Court, London, SW7 2QJ, United Kingdom

British Library Cataloguing-in-Publication Data
A catalogue record for this book is available from the British Library

Additional hard and PDF copies can be obtained from orders@intechopen.com

Recent Advances in Robot Manipulators
Edited by Serdar Küçük
p. cm.
Print ISBN 978-1-80356-266-7
Online ISBN 978-1-80356-267-4
eBook (PDF) ISBN 978-1-80356-268-1

Meet the editor

Serdar Küçük received a BA and an MSc from Marmara University, Istanbul, Turkey, in 1995 and 1998, respectively. He received a Ph.D. from Kocaeli University, Turkey, in 2004, where he is currently a full professor in the Department of Biomedical Engineering. He has several scientific publications to his credit, including international conference papers, journal papers, books, and book chapters. He serves as a reviewer of several well-known robotic journals. He is also an editor of scientific books. His research interests include optimization, control, and kinematics and dynamics modelling of serial and parallel robotic manipulators. Lately, Dr. Küçük has also been interested in designing electrical controlled, above-knee prosthetics and hand–wrist rehabilitation robots, surgical robots, and biomedical robotic devices.

Contents

Preface

Robot technology has reached a very interesting level, especially in terms of military robots, medical robots, courier robots, autonomous security robots, and robots in agriculture and entertainment. Soon, it seems inevitable that robots will replace many scientists and workers. While all this is going on, it is obvious that mass robot design and production will only be successful with a dedicated high-level education system conducted in universities. The name of this education system is robotic engineering. Talented and innovative robotic engineers who design competitive and advanced technological robotic systems can only be trained under this high-level robotic engineering system.

Every passing year, many new robotic devices (such as robot dog Spot, kangaroo robots, robot birds, humanoid robot Asimo, robot Cheetah, and the Vinci Surgical System) have been designed and produced for different purposes. These new robotics systems possess very advanced control systems and computer algorithms that will cause significant changes in production systems and lifestyles. It is clear that these new robotic systems have started a trend, attracting young people to work in this field. This trend is a sign that the future of robotic engineering will be very bright.

This book explores the latest developments in robotics as well as current robotic systems and applications. I would like to thank all authors who contributed to this book with their valuable novel ideas and their knowledge of current developments in robotics.

Serdar Küçük, Ph.D.
Full Professor,
Technology Faculty,
Department of Biomedical Engineering,
Kocaeli University,
Turkey

Section 1

New Trends in Robot Engineering

Chapter 1

Introductory Chapter: On the Verge of a New Age – The Age of Robotic Engineering

Serdar Küçük

1. Introduction

As a result of the industrial revolution, machines began to enter our lives steadily. Later, the acceleration of mechanization caused technological tools to become even more capable. Robots, which were produced only as toys before, started to show themselves in business life in the twentieth century. This robotic technology, which has spread to the world in a short time, has started to be used in many areas from cargo transportation to car painting, from electronic card production to material selection, and from surgery to rehabilitation. With the development of technologies used in this field, robots have come to such a point that they have already opened the doors to the fantastic world of the future. In order to understand the point where robots have come, let us briefly summarize some important robotic products that have been developed recently.

The robot dog named Spot, developed by Boston dynamics, can walk on any terrain. Spot is able to continue walking while maintaining its balance even in collisions [1]. Another robot named Handle, developed by Boston dynamics, can accelerate and slow down in a very short time by controlling its wheels well [2]. Examining the anatomical structure of a kangaroo, Festo's engineers designed and produced a jumping kangaroo robot weighing 7 kg and measuring 1 m in length. This kangaroo can jump forward 40 cm vertically and 80 cm horizontally [3]. The Smart Bird is an ornithopter developed by Festo's Bionic Learning Network based on the herring gull. This robotic bird has a mass of 450 g and a wingspan of 1.96 m. In April 2011, Smart Bird was introduced at the Hannover Fair [4]. The most distinctive difference of this robotic bird from its predecessors is that it can take off for itself and land after flying. Cassie, which is thought to be used for search and rescue, cargo transportation and military purposes, was developed by Agility robotics [5]. First unveiled in 1986, Asimo was developed mainly to help people. The world's most advanced humanoid robot Asimo, which was developed by Honda motor with a weight of 50 kg and a length of 130 cm, can walk on two legs at a speed of 6 km/h [6]. The robot Cheetah 3, produced by MIT, can successfully perform many movements that can make even real cheetahs jealous. The cheetah, weighing 90 kg, can jump, climb stairs without seeing it, reach a speed of 50 km/h, and regain its balance in sudden jolts [7]. The Vinci Surgical System is the most widely used robotic system in hospitals. In 2000, it becomes the first commercially available robotic surgical system in the United States. The Vinci surgical platform has been used in several operations like gynaecologic surgery, cardiac surgery, thoracic surgery, and urologic surgery [8]. The robotic systems mentioned above are changing the way of life of humanity gradually. Today, robots now work in the same office as humans and compete with humans to have many jobs. It seems unavoidable that people will leave some of the work to robots in the near future [9–11]. Although it has not been named yet, the name of the age we live in is the age of robotic engineering.

Thousands of engineers are working non-stop to develop the above amazing robotic systems. For example, mechanical engineers spend hours on the structural design of these robotic systems. In addition, they draw the kinematic [12] and dynamic [13] equations of these robotic mechanisms and perform stiffness analyses [14]. Electrical engineers are working on the selection of the most suitable actuators and sensors for these robotic systems. Electronics engineers spend time designing the electronic boards required for these robotic systems to operate at maximum performance. Computer engineers, on the other hand, spend time on the development of the necessary software in order for these robotic systems to move smoothly in the desired trajectory.

Apart from these main subjects, many engineers in other fields of science (such as materials engineering and mathematical engineering) spend a lot of time in order for these robot systems to work smoothly. Bringing together thousands of engineers from many different disciplines to manufacture such amazing scientific robotic systems has become a more difficult task in today's world than it used to be. Moreover, it has become more expensive than before to design and manufacture these magnificent robotic systems with many employees in different cities and even countries, especially by bringing together thousands of parts and assembling them into a single product.

2. Conclusions

It seems unlikely that engineers from many different disciplines described above will come together to design and produce future's competitive and advanced technological robotic systems. Having an advanced technological manufacturing system requires a more compact business environment where laboratories, workshops and engineers are gathered under one roof. At this point, it can be said that mechatronics engineering already has such a competitive structure. It should be noted that mechatronics engineering is not just a discipline in which robotic systems are designed solely. Apart from robotic systems, mechatronics engineering is a discipline that designs automation systems in many different fields of industry. Robotic systems are only a small part of this wide spectrum of mechatronic engineering. In addition, mechatronics engineering is generally concerned with the automation of industrial robots in different business areas. However, in the future's world, besides industrial robots, robot production is needed for many different fields such as medicine (medical robots), military (military robots), transportation (courier robots), human security (autonomous security robots), agriculture, entertainment, space and underwater exploration. Therefore, there is a need for a more specific field of science. Even if this branch of science has not been named yet, it is robot engineering. In order to design competitive and advanced technological robotic systems, we need to train talented and innovative robotic engineers. The way to do this is through the planning of an advanced robotics education system. The institutions where this advanced robotic education system will be implemented are universities. The countries that are already planning the establishment of these schools will be the leading countries. Since these

DOI: http://dx.doi.org/10.5772/intechopen.107072

countries will manage the field of robotics engineering, they will be the countries that benefit the most from the trade of this field. While this trade was initially at the level of millions of dollars, it is thought that it will reach the levels of tens of billions of dollars in a very short time. As a result, it is thought that the impact of robotic engineering will cause significant changes in production systems and lifestyles. In the next book, discussions will be made about how this robotics engineering education system should be.

Author details

Serdar Küçük
Technology Faculty, Department of Biomedical Engineering, Kocaeli University, Kocaeli, Turkey

*Address all correspondence to: skucuk@kocaeli.edu.tr

References

[1] The Robot Dog Spot. United States: Boston Dynamics; 2016

[2] Handle. United States: Boston Dynamics; 2019

[3] Bionic Kangaroo. Festo, Ruiter Straße 82 73734 Esslingen Germany: 2014

[4] Smart Bird Animation. Festo. 2011

[5] Cassie. United States: Agility Robotics and Oregon State University; 2016

[6] Asimo. Honda. Available from: http://asimo.honda.com/

[7] Bledt GP, Katz M, Carlo B, Wensing J, Kim P. MIT Cheetah 3: Design and Control of a Robust, Dynamic Quadruped Robot. 2018

[8] Dwivedi J, Mahgoub I. Robotic surgery—A review on recent advances in surgical robotic systems. In: Florida Conference on Recent Advances in Robotics; Boca Raton, Florida. 2012

[9] Küçük Ö. The new social politics notions, the changing from working power accumulation to the safety to live. Akademik Hassasiyetler. 2019;**6**:25-60

[10] Küçük Ö. Emeğin Dijital Dönüşümü ve İstihdam. International Journal of Human Studies. 2021;**4**:307-328

[11] Küçük Ö. Industry 4.0, artificial intelligence and new period of labour relations. In: 10th International Symposium on Intelligent Manufacturing and Service System. Sakarya/Turkey: Sakaraya University; 2019. pp. 60-68

[12] Kucuk S. Dexterous workspace optimization for a new hybrid parallel robot manipulator. Journal of Mechanisms and Robotics. 2018;**10**(6):1-8

[13] Tsai LW. Solving the inverse dynamics of a Stewart–Gough manipulator by the principle of virtual work. Journal of Mechanical Design. 2000, 2000;**122**(1):3-9

[14] Gosselin C. Stiffness mapping for parallel manipulators. IEEE Transactions on Robotics and Automation. 1990;**6**(3):377-382

Chapter 2

New Trends in Robots Engineering with Professional Software SolidWorks

Ciprian Dragne, Isabela Todirite, Marius Pandelea, Corina Radu Frenț, Petru-Alexandru Cotfas, Veturia Chiroiu and Mihaiela Iliescu

Abstract

Engineering robotic systems stand for a challenging complex process, closely related to product development phases. Society's needs and requirements generate the idea for new robot products, which are sketched as an initial concept. This is the moment when the design phases start, engineers continue their work by evaluating and optimizing the mechanical parts according to many criteria: kinematics, dynamics, the strength of materials, NVH, thermal assessments, etc. Finally, there are established specifications for prototype execution, environment sustainability, end-user specifications, and recycling requirements. All these phases could be implemented into smart software. SolidWorks is such software enabling the creation of new mechanical designs automatically based on its programming tools. This chapter is focused on relevant advanced capabilities of SolidWorks software to assist engineers in achieving a new advanced level in mechanical design, that of automatically generating new or modifying existing concepts according to the requirements. By using professional software in research studies, new engineering procedures can be developed in order to automate the concept and design phases for many concurrent engineering methodologies, design optimization methods, manufacturing, documentation, or end-user specification. Case studies on the different types of robot systems used in healthcare and assisted living are presented.

Keywords: robots, mechatronic system, optimization, concurrent engineering, Solidworks, healthcare, assisted living

1. Introduction

In any concept of a robotic system, the start is with a vision of the product, and an idea to design. As soon as the work for it begins, the project outline and preliminary calculations to evaluate the concept are to be done. These could involve calculi for evaluating the kinematic performance, the working space, evaluating the loads during operation, evaluating the type of materials used, calculation of components' pre-sizing after a few significant simulation cases, calculations of the product life and cost calculations [1].

After these preliminary phases, next step is that of detailed design. Building of a detailed 3D model, kinematic evaluation of the system, strength calculations for all required cases, evaluation calculations according to other criteria, steps for optimizing components that do not meet the needed requirements, durability calculations, and service life assessment of components - all of these represent bases of a complex engineering process [2].

Both preliminary and detailed calculation cases are multidisciplinary assessments. These may include both strength analysis and assessments of dynamic behavior, thermal effect, interaction with substances that require an assessment of fluid flow, etc.).

This chapter, gives an overview of some practical methods of product design, following all the aspects mentioned above, but using modern methods and a professional CAD program. The use of modern methods means the use of personal computers both for the evaluation by already set methods and for the development of new methods, which bring to a new light all the features evaluated by older methods [3].

Finally, it is to be enlightened that the use of advanced CAD programs could enable, even "extravagant" facility such as the automatic realization of components, the evaluation of many features under the perceptions of competitive engineering, and even the automatic elaboration of projects, drawings, and technical specifications of a product [4].

Concurrent engineering is a method of designing, evaluating, and developing a product, in which the various stages of its evolution are solved simultaneously, rather than iteratively. This method reduces the time of design, implementation in production, the time required to launch onto the market, etc., which leads to improved productivity and reduce costs [5, 6].

Mechanical design of mechatronic system's components, modeling and simulation for further validation are presented next, highlighting the most efficient use of design, analysis, and manufacturing tools offered to users by SOLIDWORKS.

2. Current state in design

In the early days of engineering, designers used empirical, sequential, uni-criteria methods. All those methods were executed most of them on paper, and the project files included explanations of all the required phases and calculation steps. The need for faster evaluation methods has even necessitated the advent of calculating machines. At first, computers were not programmable, the commands were executed line by line, and so were the results. With the advent of programmable computing tools, the earlier stages of product development have been completed more quickly and new methods have been developed and diversified. Thus appeared the first dedicated computing programs. In fact, the first programs were developed even for engineering purpose of calculating some parameters that required long time of information processing. One of these areas was the methodology for calculating a trajectory required for a space shuttle, taking into account all the factors that influence this trajectory [7–10].

Over time, software for complex mathematical calculations have emerged. Some of these programs are MATLAB and Mathematica. These allow to introduce command sequences to evaluate mathematical expressions.

At the beginning, most function evaluations were performed in simplified representation systems, with variations of curves only in plane. Thus appeared the first CAD programs (for the graphical representation of the components) but they showed the pieces only in 2D representation.

However, the components encountered in reality are three-dimensional. The whole visible universe unfolds in a huge 3D scene. The emergence of components representation as three-dimensional structures seems a natural step in the development of design software.

According to the main purpose, the design phases of a product can be divided into:

- evaluation phases of the assembly of parts and subassemblies;
- phases for analyzing the stress and strain state of the components or, the strengths of the whole assembly (as a structure);
- phases of making the documentation;
- execution phases of the parts;
- execution of prototype;
- phases of experimental tests;
- phases of entry into production.

Each of the above phases has acquired well-known names over time, as mentioned next.

- Modeling the geometry and the way of assembling the system components has been called CAD (computer-aided design);
- Carrying out the structural evaluation phases and the characteristics of the product has been called CAE (computer-aided engineering);
- Carrying out the execution phases of the product has been called CAM (computer-aided manufacturing).

If in the beginning, CAD-CAE-CAM programs were clearly differentiated from each other, in recent years there has been a trend of unification, so software developers have begun to unite to bring out more and more high-performance engineering applications. CAD developers today seek to benefit from the performance of subassembly calculation and evaluation programs, while CAE application developers seek to benefit from the performance of CAD applications for viewing complex 3D assemblies. One of the three areas has lagged behind in recent years. This is the realm of CAM applications. It was only after the advent of CNCs that a clear connection with the above applications was achieved. The ultimate goal of any product is its physical realization. It is obvious that the methods of physical realization can influence all the previous methods of design, going back even to the initial conception and idea phase.

The most advanced 3D geometry modeling applications today are Catia, Solidworks, Pro-Engineer, Inventor, and Ansys. There are also others, many are developed ad hoc in various universities around the world.

As mentioned above, the methods for evaluating the performance of systems were initially developed separately from the CAD part, although applications of materials strength methods have gone hand in hand with engineering from the beginning. In any kind of concept, the engineer had to have knowledge about the possible tasks that could be supported by the designed structure. A very common method used in recent years to evaluate the performance and mechanical properties of a structure is the finite element method (FEA). The most advanced professional applications in these methods are Abaqus, Ansys, Nastran, Pam-crash, Ls-Dyna, Solidworks, Comsol, Autodesk Simulation, etc. [11–16].

A special field of research derived from evaluating the performance of a system is studying its kinematics. Establishing and evaluating the performance of a system in all positions of the workspace, including the determination of displacements, speeds, and accelerations in the system, proved necessary in the study of its dynamics. In the case of systems where the components have relative movements with respect to each other (complete rotations, complex trajectories, necessary speeds and accelerations as tasks, etc.), specific applications have also appeared for the evaluation of their kinematics. Relevant ones, as best performing applications, are Adams and Solidworks.

The development of ideas and concepts, the assembly of designed components, the evaluation of performance of the whole system, the optimization of the proposed solutions, and the choice of the best manufacturing method could be achieved with the help of modern mechanical design tools. Engineers, assisted by professional CAD-CAE-CAM design applications, could be involved in the entire product development process, from concept phase to prototyping, structural strength assessment, dynamic behavior assessment, effective manufacturing by choosing execution methods of physical and final assembly, as well as completion of all product documentation including maintenance and recycling phases.

3. Concept development

Basic aspects of concept development for different robot/mechatronic systems are presented next.

3.1 Concept of robot for laparoscopic surgery

This subchapter presents the concept of a surgical robot, designed in a kinematic chain configuration with parallel components. The medical purpose is that of surgical robot usable in brachytherapy procedures.

The development of the concept from the idea to the detailed CAD model for the surgical robotic system went through the following stages:

- establishing the preliminary dimensions;
- establishing the first configurations;

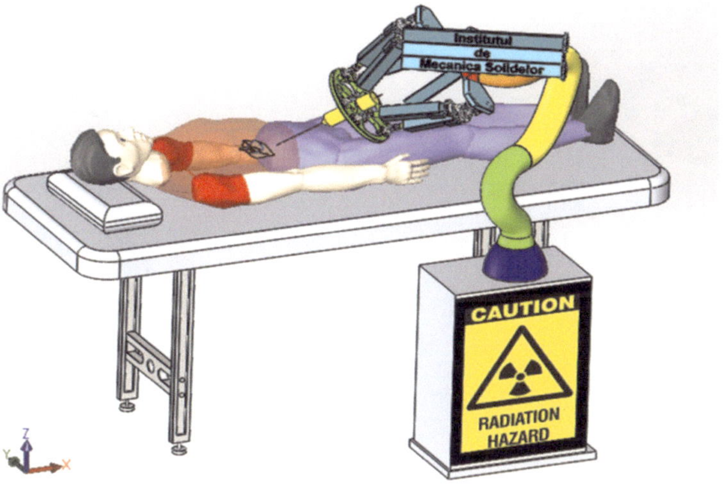

Figure 1.
Surgical robot - CAD model.

- optimization of the first configurations for kinematic performance parameters, the workspace, the optimal trajectories, the kinetic parameters, etc.;
- selection of components for acquisition, actuators type;
- structural optimization based on the criteria of mechanical strength, selection of the best materials, assessment of NVH behavior and thermal effects, etc.;
- establishing the optimal locations for the acceleration sensors intended to monitor the structure;
- evaluation of the product life;
- mechatronic system control strategy - choosing the programming platform for control functions;
- choosing the built strategies of the parts;
- evaluating the costs of the product.

The (up to date) model of the robot for laparoscopic surgery is presented in **Figure 1**.

3.2 Concept of mechatronic system for visually impaired people

This subchapter presents the concept of a modular mechatronic system for visually impaired people. The idea of the concept came after studying the assistive devices present on the market, and the conclusion that they can be improved. This concept aims both to address the issue of assistive devices for the visually impaired and to present a virtual prototype of a modular mechatronic system for visually impaired

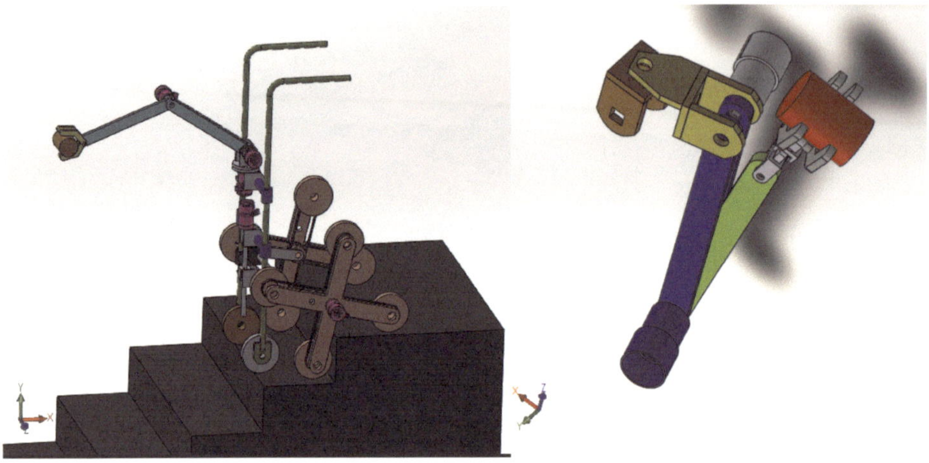

(a) (b)

Figure 2.
Visualy impaired mechatronic system - CAD model. (a) whole mechatronic system (b) robotic arm sub-system.

people who have acquired other mild or medium deficiencies during life or even from birth. The virtual prototype is not a prototype for a conventional device, but a custom one consisting of modules that can be added as the person has adapted to the basic set and other needs have been identified. The prototype of the mechatronic system contains multiple modular systems (location, color identification, object bypass, haptic feedback - audio, and others). People with visual impairments, in addition to having to move in closed or open spaces, need to carry out daily activities which, in most cases, require the recognition of objects in the environment.

The mechatronic system is designed so that people with mild neuro-motor dysfunctions could also be helped to move, to go up and down stairs and the robotic arm to help the blind person when there is a window on the taxiway, as well as doors from open cabinets to push them.

This system is intended to help the visually impaired in difficult times for them, such as bypassing obstacles in the way. The user in the situation when will use it in open spaces will face all kinds of situations that he has to manage.

From the mechatronic point of view, the system's basic components are (see **Figure 2**):

- the motion system – with wheels enabling climbing up/down stairs;
- the robotic arm system - will push window handles, open cabinet doors that stand in the way of the subject, grab objects from shelves, etc.

3.3 Concept of anthropomorphic walking robot foot

Anthropomorphic walking robots are extremely complex systems whose main problem is static and dynamic stability in the unknown environment. In most cases, the fulfillment on the tasks depends on:

- mobility of mechanical structures;

- identification and recognition of objects and obstacles;
- navigation in the workspace.

Figure 3 (a) shows an innovative 3D solution in terms of mechanical structure, which is generated by the SolidWorks software. The idea was to design a walking subsystem for anthropomorphic robots that provides increased mobility and energy efficiency for effectors movements. It was chosen the version with sole consisting of three segments to ensure the extra force of movement through the toes and articulated heel for increased cushioning on contact with the support surface. The building blocks, both mechanical and electronic, are mostly chosen from SolidWorks' database.

Figure 3 (b) shows the 3D solution of the anthropomorphic robot sole with articulated heel generated by the SolidWorks software.

The dimensions and characteristics of the designed mechanical elements are component parts of the control law that guide the heel when it lands on the support surface.

The physical parameter defines the rotation of the heel for shock cushioning is as follows (Eq. (1)):

$$I_p = I_e \times \left[1 + F(t_i) \times \mathcal{S} / \left(k_m^2 + k_g^2\right)^{1/2}\right] \tag{1}$$

where I_p is the weighted inclination of the normal point of contact of the heel relative to the vertical position, I_e is the estimated inclination by calculation of the normal point of contact of the heel relative to the vertical position, $F(t_i)$ is the force on

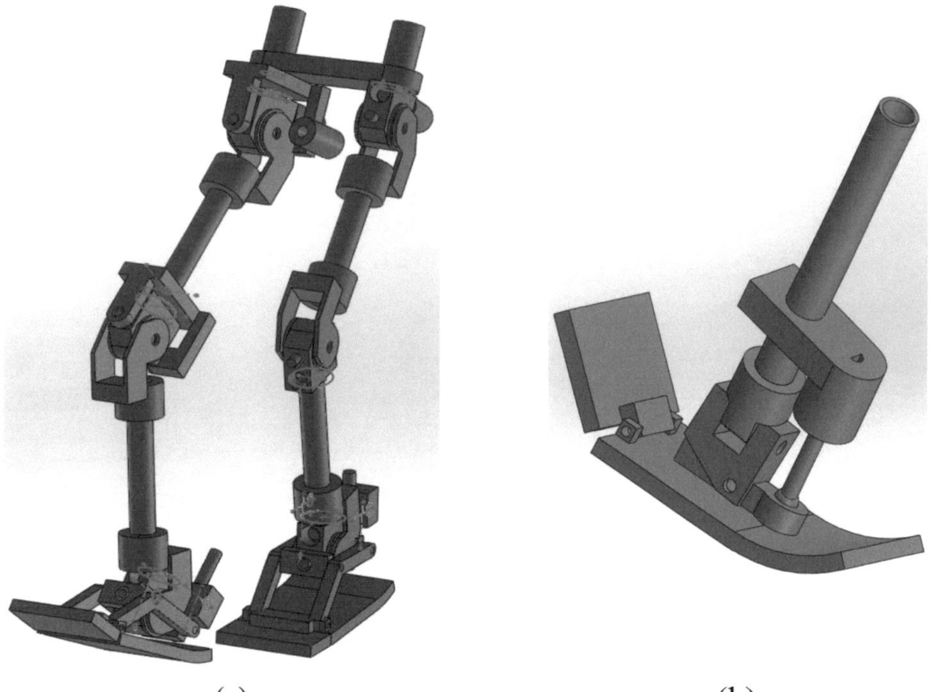

(a) (b)

Figure 3.
Mechatronic system for antrophomorphic robot foot. (a) General scheme (b) Sole detail.

the support surface at the moment t_i, δ is the damping displacement, k_m and k_g being the rigidities of the material, the heel of the foot and support surface, respectively.

From the point of view of the motors used, we chose the brushless DC version, in order to have a fast, precise movement and a clearly superior control of the position of the robot's legs segments.

3.4 Concept of mechatronic system for locomotors disabled people

The mechatronic system's concept presented in this subchapter is aimed to help people with disabilities, people who present different forms of paralysis of the lower limbs, or simply those who are in a period of medical recovery after undergoing operations that restrict their mobility. By the system (see **Figure 4**), these people are assisted in in transferring from the wheelchair to a vehicle or in performing various household activities.

The transfer from a wheelchair to a vehicle is the operation with a high degree of difficulty, because the person's loading area is limited, the movements of the system are limited and the possibility that the system or, worse, the person, will collide with car parts (door, pillar, chassis) is high. The system can be adapted on several types of vehicles depending on the car geometry (pillar shape and dimensions), as its concept is that of adjustable clamping hinge (see **Figure 5**).

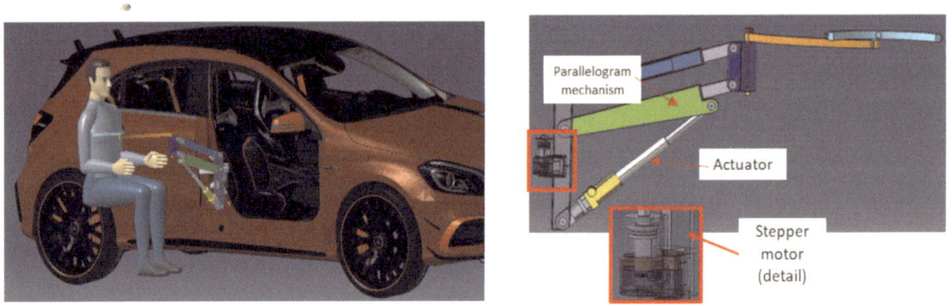

(a) (b)

Figure 4.
Mechatronic system for locomotors disabled. (a) transferring to car front seat (b) paralellogram system.

Figure 5.
Adjustable clamping hinge.

4. Case studies for robots systems engineered assisted by professional SolidWorks

Robots and mechatronic systems' engineering using professional SolidWorks is relevant in any stage of new product development. Some case studies for the designed systems are shown next.

4.1 Analysis of robot for laparoscopic surgery

A surgical robot needs very well-prepared documentation that includes many chapters about:

- the medical procedures considered;
- the main requirements of the medical robot: workspace, used tools;
- the mechanical characteristics required or to be taken into account;
- kinematic, dynamic, medical imaging performances, etc.;
- materials used, medical instruments;
- the maintenance specification of a product that works in the medical field;
- recycling conditions.

4.1.1 Workspace

The workspace is the entire space around the "home" position where the robot can move. This is the robot workspace. The evaluation of the workspace is done by knowing the direct kinematics equations for a robot with configuration based on parallel structures (see **Figure 6**). Such workspace is presented in gray color in **Figure 6**-b with a section in red color. But near this workspace should be added the

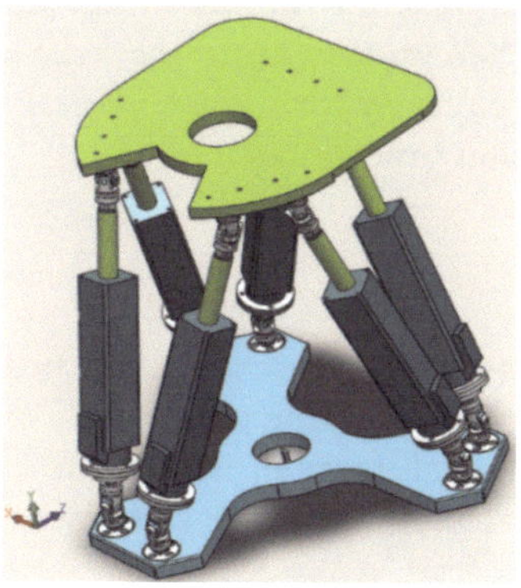

(a)

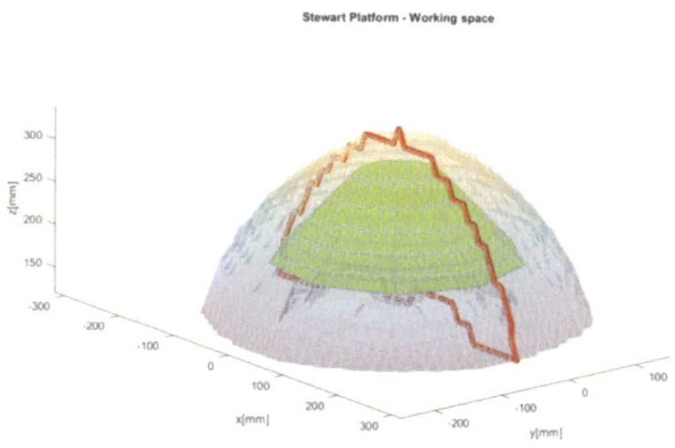

(b)

Figure 6.
Workspace of hexapod robot. (a) hexapod robot (b) robot and required workspace.

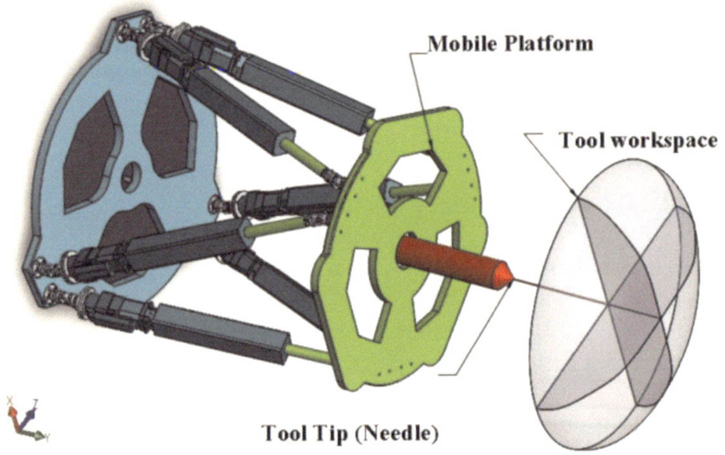

Figure 7.
Robot workspace evaluated with Solidworks API.

required space on design. This space is presented in green color, which is a special workspace in the sense that its definition is made in cylindrical coordinates. This type of workspace has a number of advantages in terms of solution symmetry and the disadvantage that the implementation of its equations is complex.

The workspaces above were evaluated using the MATLAB program based on a parametrization in the position of the platform using only generalized displacement at its center. These mathematical equations can also be implemented in a CAD program in order to benefit from all its graphical performance. A such example is in **Figure 7**, which shows the workspace of a hexapod robot evaluated using procedures developed with the Solidworks API. The working method uses the procedures presented in [17] to find the boundary of the workspace at the operating limit of the actuators.

API (Application Programming Interface) is a background support program from Solidworks CAD that allows programming commands to be written in a specific language to automate model execution based on procedures implemented internally in software.

The methods presented for a robot's workspace do not take into account the possible collision between the elements of the system when positioning in workspace. Because the actual geometry of the components can vary greatly from the idealized shape, then it is necessary to implement collision detection procedures.

4.1.2 Collision detection

A robot is generally a machine designed to perform tasks automatically, with speed and accuracy. Even when a robot is operating properly, it may collide with people or objects that enter its workspace and may even cause personal injury or damage to those objects. A particular case is the self-collision when the robot strikes its own components from which it is built [18, 19].

Injuries or damage to others (humans or mechanical components) due to the robot's activities are classified as contact damage. The main aspect of the impact between two or more components is collision detection. How collision assessment depends mainly on the geometry of the parts, the more complicated are the components, then more

Figure 8.
A short presentation of decomposition techniques used in collision detection [20].

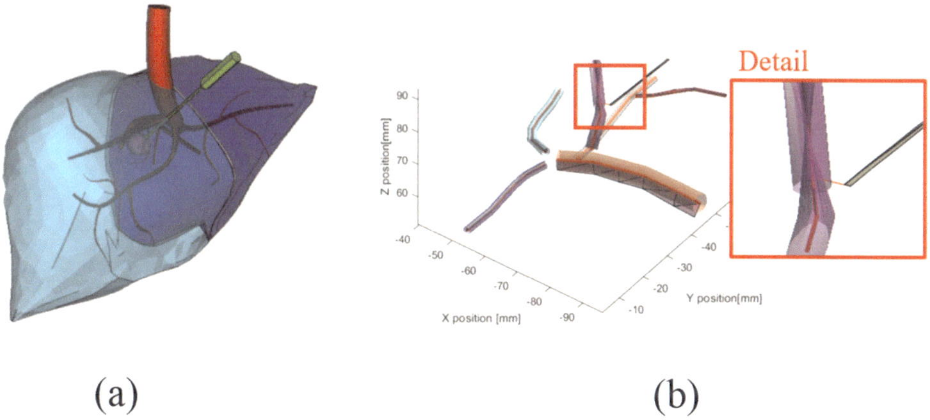

Figure 9.
Collision detection strategy at forbiden zones. (a) virtual reprezentation of liver (b) veno-arterial tree in tubes geometric primitives.

time it takes to assess possible collisions. That's why it's a good idea to start with a simplified geometry. **Figure 8** summarizes the geometry decomposition techniques used in collision detection. In paper [20], the authors presented some innovative techniques for the detection of collisions between components with complex geometry.

If the assessment of the collision is done with a veno-arterial network then the detection becomes even more complicated. For the speed of the calculations, it is recommended to simplify an artery tree until to the level of geometric primitives (**Figure 9**).

The Solidworks software is useful in collision detection with its methods to evaluate the distance between components.

4.1.3 Kinematic performances

Evaluating the configuration of a robot involves a study of choosing the best features (dimensions, types of links, type of drives, etc.) according to different criteria, such as:

- compatibility with the surgery room and with medical procedures;
- compatibility between the tools used and the workspace;
- compatibility with the kinematic performances of the robot.

The optimization of the different quality parameters of the robot must be done according to several kinematic performance criteria (see **Table 1**): singularities, determinants of the inverse of the Jacobian matrix, dexterity, global conditioning index, local conditioning index, manipulability, etc. [21, 22].

Optimization of configuration (see **Figure 10**) involves the knowledge and use of advanced methods of mathematical calculation: multi-criteria methods, genetic algorithms (GA), and matrix processing. **Figure 10**-b shows the dispersion of intermediate results when using a GA to optimize the configuration of a hexapod robot until the optimal position is found.

The kinematic evaluation of the mechanisms can also be done using a professional CAD program. Solidworks provides tools for such assessments. **Figure 11** shows the final results of the positions of a surgical robot using the Solidworks software for both the forward and inverse kinematics. **Figure 11**-a shows the results of forwarding kinematics procedures by increasing the length only at one actuator, and **Figure 11**-b shows the results of inverse kinematics procedures to put the tools at a specific

The determinant of the Jacobian matrix	$\det\left([J]^{-1}\right)$
Dexterity	$[DEX] = \left\|\det\left(J^{-1}\right)\right\| \bullet \|J\|$
Local conditional index	$LCI = \frac{\sigma_{\max}(J)}{\sigma_{\min}(J)}$
Manipulabiliy	$Manip = \sqrt{\det(J * tr(J))}$
Global conditional index	$\eta = \left(\int_{WS} k \bullet dW\right) / \left(\int_{WS} dW\right)$

Table 1.
Criteria for kinematic performances.

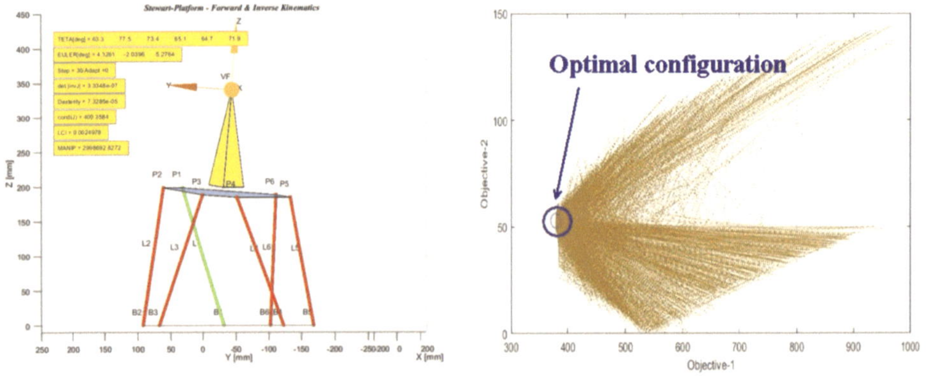

(a) (b)

Figure 10.
Assessment and optimization of kinematics. (a) assessment of configuration in real time (b) optimization using GA.

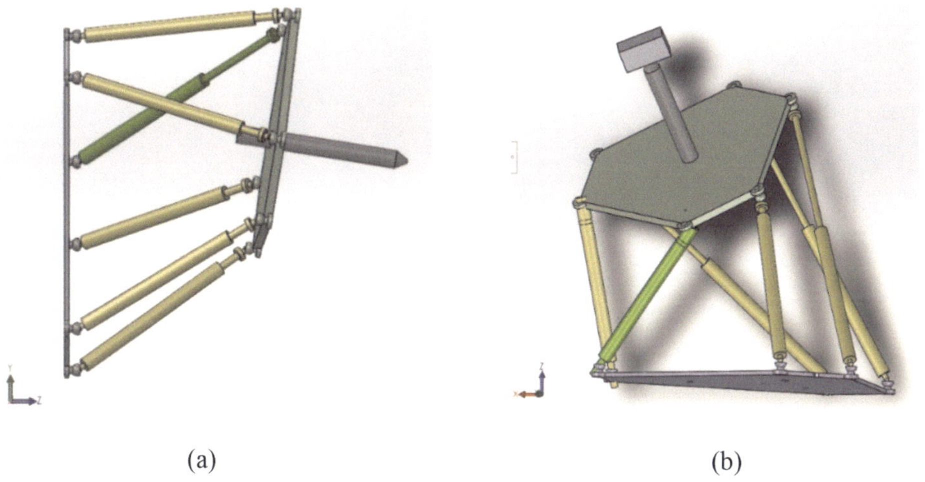

Figure 11.
Assessment of kinematics using Solidworks. (a) forward kinematics (b) inverse kinematics.

location in a specific direction. However, the methodology requires knowledge of the Solidworks API script.

4.1.4 Durability assessment

Effective service life is determined by several factors, including:

- load cycles;
- environmental factors: temperature, humidity, corrosion, etc.;
- selected materials;
- electromagnetic fields under which the system works.

Research results on evaluation of the effect certain expected loads on the service life of a product do have. A load cycle was considered. The stress level is VonMises stress extracted from 100 load cases at the same probe when the loads vary in different directions and amplitude. An example of VonMises stress level at probe for load case (no. 55) with 1000 N at tool is presented in **Figure 12**. The probe is located at platform because this part is the main for structural design optimization.

Material for platform is considered in this phase the polycarbonate plastic material. This material could be changed during further optimization steps.

The assessment of the endurance by effect of the stress state on the material took into account the Wohler curve of the polycarbonate plastic by applying the rainflow method. The results are shown in **Figures 13** and **14**. For the above load cycle the damage is 1.0434e-6 for a single stress sequence. This means that the life of the verified component is 958,405 cycles.

Rainflow algorithm is the most popular counting method used in fatigue and failure analysis for lifetime estimation of mechanical parts. The rainflow counting technique was introduced by Matsuishi and Endo in 1968 to extract closed loading

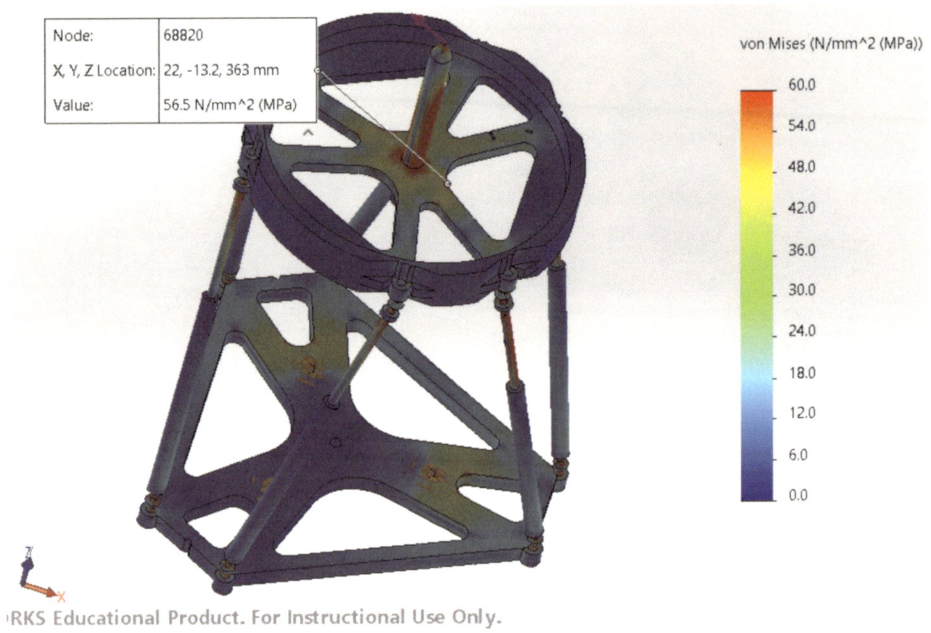

Figure 12.
Von Mises stress level at probe for load case (no. 55).

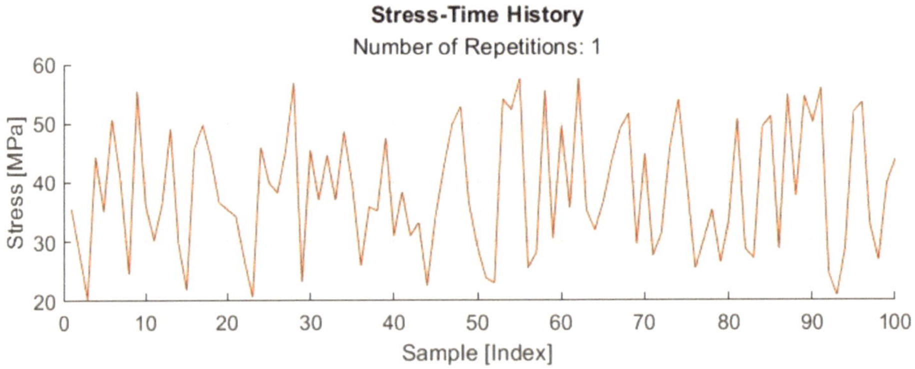

Figure 13.
Reference load cycle.

reversals or cycles for a correct estimation of fatigue. The "rainflow" was named in comparison to the flow of rain falling on a pagoda roof [23, 24]. For the studied example, the detailed rainflow diagram is presented in **Figure 15**.

Endurance evaluation can be done also using Solidworks software, but not with such detailed diagrams for post-processing and reports but with something else which is also important – durability in 3D fields.

4.1.5 Thermo-mechanical evaluation

Thermal flow can influence the performance of a robot. These aspects are even more important in the case of a surgical robot. The results of thermo-mechanical

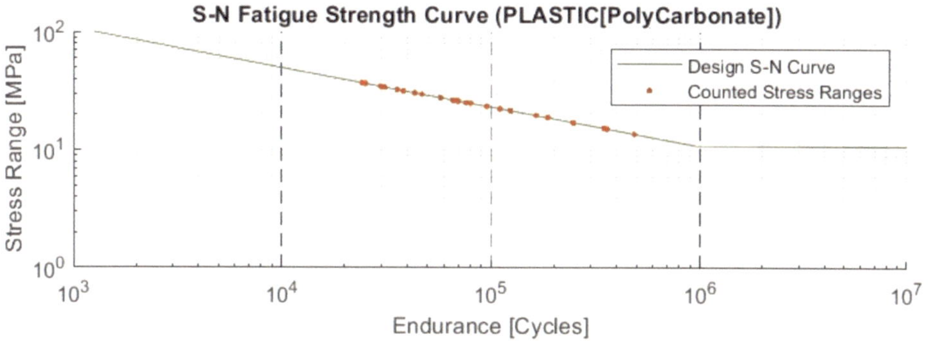

Figure 14.
Fatigue curve for plastic -polycarbonate.

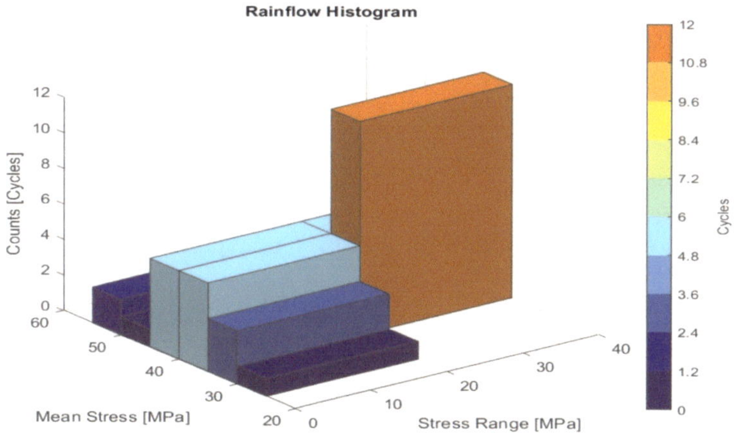

Figure 15.
Rainflow diagram.

simulation for the surgical applications hexapod robot are further presented. Thermal load is considered overheating of an actuator from 18 to 30 degrees Celsius because of a failure at its motor. Because of to this thermal load, the total deviation at the end of the tool is 0.25 mm for presented model (see **Figure 16**).

In a surgical room, there could exist many thermal sources from the heating of electrical systems, heating flow, radiation, convection, advection, etc. An evaluation of thermal effects is mandatory for a surgical robot.

4.1.6 Damage detection

Structural Health Monitoring (SHM) is applied today to mechanical structures that require significant costs, for structures difficult to inspect or where human safety is a priority. The main task in this subject is damage detection. But damage detection involves today also methods to estimate damage location, damage size, or other additional information about the damaged area.

In this sub-chapter is presented an improved DLAC method for damage localization technique applied to a surgery robot structure. Basically, method use frequency

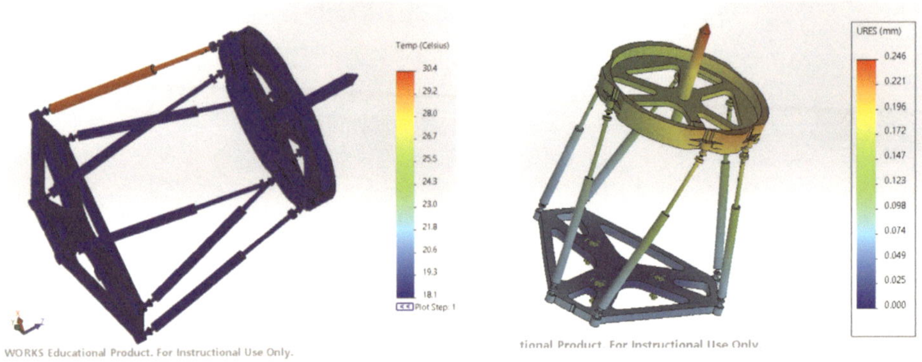

(a) (b)

Figure 16.
Thermo-mechanical analysis. (a) thermal field (b) thermal deformation.

shift for damage detection (Eq. (2)). The DLAC criterion was improved by transform equation into a probability index for a better assessment (Eq. (3)).

$$DLAC(i) = \frac{\left|\Delta\omega_E^T \Delta\omega_A(i)\right|^2}{\Delta\omega_E^T \Delta\omega_E \left(\Delta\omega_A^T(i) \Delta\omega_A(i)\right)} \tag{2}$$

where $\Delta\omega_X$ is the frequency shift for analytical or experimental (A, E) model.

$$probability(DLAC) = RESCALE(DLAC, 0, 100) \tag{3}$$

Successful damage localization technique depends on eigenmodes number. In **Figure 17** there are presented the set damages locations and final diagram for DLAC probability index.

The probability index of DLAC criterion shows the maximum value exactly where the damage was imposed in each model. This is possible when exist only one damaged area. In case of simultaneous existence of several damages, the DLAC method is no longer effective, being necessary to corroborate with other criteria for a correct evaluation.

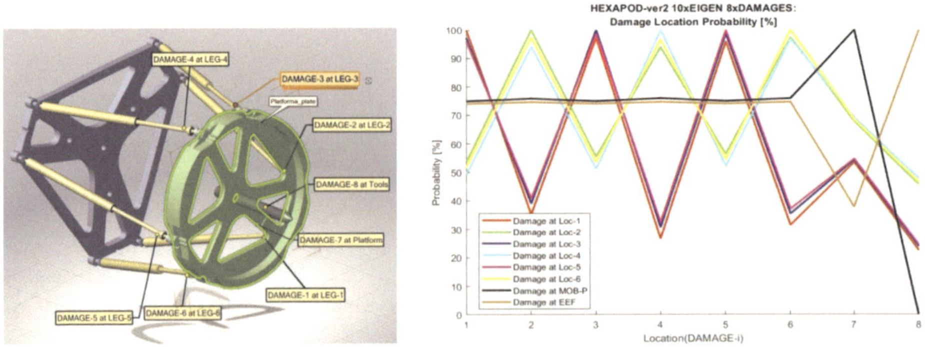

(a) (b)

Figure 17.
Damage localization using 10 eigenmodes. (a) 8 damaged locations (b) DLAC probability index assessment.

Detailed examples for damage detection probability index can be found in [25].

The complete task of the simulations for surgical robot was developed based on MATLAB software [26], SOLIDWORKS Educational [27] and user defined programming routines, VB, API, etc., [28].

4.2 Damage detection for robotic arm

A similar damage localization technique was tested on the robotic arm of the visually impaired mechatronic system (see **Figure 2**. (a)). The simulation uses 8 eigenmodes and 6 damages.

One can conclude that simulation results prove the estimation for the damage location, as the higher probability of damages resulted for location no. 5 (see **Figure 18**).

4.3 Optimization of geometric configuration for a hand prosthesis finger

The model of finger hand prosthesis has been built, as part of a research on biomechanical prosthesis (see **Figure 19**) and one important step prior to prototyping was that of optimization its geometry, specially its mechanical components (levers) dimensions.

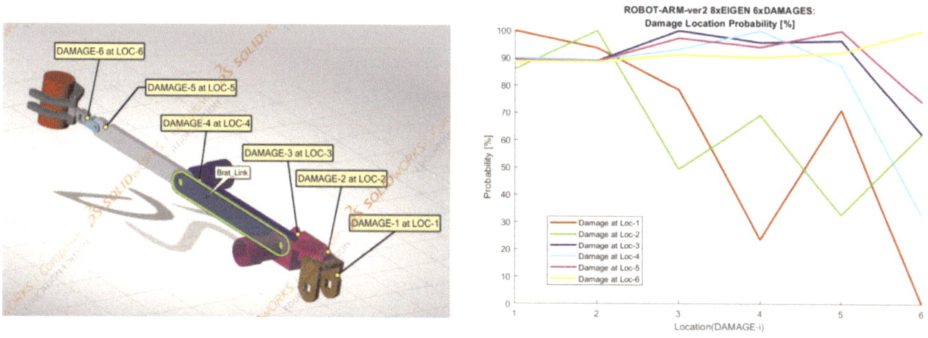

(a) (b)

Figure 18.
Damage localization using 8 eigenmodes. (a) serial robot - 6 damages (b) DLAC probability index.

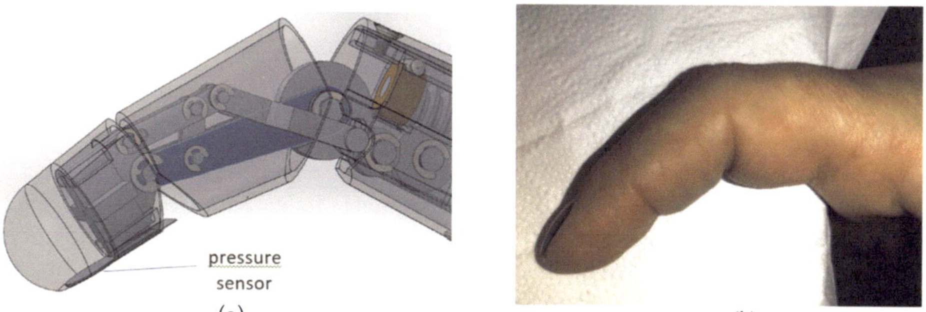

(a) (b)

Figure 19.
Hand finger's motion. (a) finger model (b) real finger.

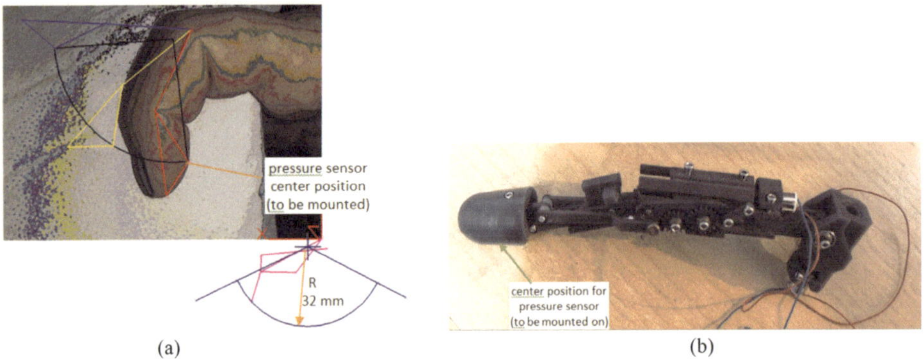

Figure 20.
Hand finger components. (a) tip finger trajectory (b) finger prototype.

Basic finger's motion is that of rotation of phalanges so that to grab an object. Based on the 3D model, the fingertip trajectory was simulated (so that to be on a circle arc) and thus the position of the pressure sensor was determined (see **Figure 20**) at a radius of 32 mm. In the 3D model, the sensor is positioned at 28 mm. So, adjustments of levers' lengths have to be done.

5. Conclusion

Modern CAD design tools bring many benefits to their users. Solidworks is a design program with many features in mechanical design and not only for mechanical domain. Originally developed for the visualization and virtual construction of complex ensembles, the software has been developed in recent years into a complex assessment of overall tolerances, parts parameterization, kinematics, structural strength, the performance of the product, cost evaluation, etc. These benefits can be multiplied by using Solidworks API programming codes.

All stages from product design study to prototyping and/or manufacturingcan be solved with a single working tool, Solidworks software.

Technical conclusions:

1. Robot surgery design involves many aspects from medical, physics, engineering, prototyping and manufacture fields [29];

2. Damage localization is a reliable technique for monitoring of mechanical structures during evaluation, testing and service;

3. Programming tools increase performances in engineering design [30–33];

SolidWorks is the most advanced tool for engineers today because of its range of applications that can be involved in real product design and manufacture Solidworks is even more advanced. Solidworks API can build assemblies in a fully automatic fashion. An entire assembly can be obtained based on the parameterization of the components and commands in the API programming code. And the adventure to new levels of engineering design just starts. Multi-physics, multi-disciplinary and multi-platform in computer and engineers'ability will grow.

The result of the design with SolidWorks software will be analyzed cinematically and dynamically in order to verify and validate the concept and solution for the robot system engineered.

Acknowledgements

We thank CADWORKS International SRL for the support and help given in this research, https://www.cadworks.ro/.

We also thank the Romanian Academy for the support of this research.

Part of this work was supported by Grant of the Romanian ministry of Research and Innovation, CCCDI–UEFISCDI, PROIECT 653 PN-III-P2-2.1-PED-2019-0085, Contract no. 447PED/2020.

Conflict of interest

The authors declare no conflict of interest.

Author details

Ciprian Dragne[1], Isabela Todirite[2], Marius Pandelea[2], Corina Radu Frenţ[2], Petru-Alexandru Cotfas[2], Veturia Chiroiu[1] and Mihaiela Iliescu[2]*

1 Department of Mechanics, Institute of Solid Mechanics - Romanian Academy, Bucharest, Romania

2 Department of Robotics and Mechatronics, Institute of Solid Mechanics - Romanian Academy, Bucharest, Romania

*Address all correspondence to: mihaiela.iliescu@imsar.ro

References

[1] Moisil G. Fizica pentru ingineri/ Physics for Engineers. Vol. 1. Bucuresti: Editura Tehnica; 1973

[2] Mortenson M. Vectors and Matrices for Geometric and 3D Modeling. U.S: Industrial Press Inc.; 2020 ISBN-10: 0831136553

[3] Jeremy Zheng L. CAD, 3D Modeling, Engineering Analysis, and Prototype Experimentation. Springer Boook: Industrial and Research Applications; 2015. ISBN: 978-3-319-05921-1

[4] Solidworks Community. https://my.solidworks.com/ [Accessed: May, 2022]

[5] Turino JL,Turino J. Managing Concurrent Engineering: Buying Time to Market: A Definitive Guide to Improved Competitiveness in Electronics Design and Manufacturing. New York: Van Nostrand Reinhold; 1992

[6] Shina SG. Successful Implementation of Concurrent Engineering Products and Processes. John Wiley & Sons; 1993

[7] Neugebauer OE. The Exact Sciences in Antiquity. Princeton: Princeton University Press; 1952. ISBN 1-56619-269-2

[8] Bell G, Newell A. Computer Structures: Readings and Examples. New York: McGraw-Hill; 1971. ISBN 0-07-004357-4

[9] Bowden BV. The language of computers. American Scientist Jr. 1970; **58**:43-53

[10] Patterson D, Hennessy J. Computer Organization and Design. San Francisco: Morgan Kaufman; 1998. ISBN 1-55860-428-6

[11] Logan DL. A first course in the finite element method. Cengage Learning. 2016. ISBN 978-0495668251

[12] Maksay IS, Bistrian DA. Introducere in metoda elementelor finite. Ed. Iaşi: Cermi; 2007. ISBN 978-973-667-324-5

[13] Bathe KJ. Finite Element Procedures. Cambridge, MA: Klaus-Jürgen Bathe; 2006 ISBN 978-0979004902

[14] Blumenfeld M. Introducere in metoda elementelor finite / Introduction to Finite Elements Method. Bucureşti: Editura Tehnica; 1995

[15] Cuteanu E, Marinov R. Metoda elementelor finite in proiectarea structurilor. Timisoara: Editura Facla; 1980

[16] Beirão da Veiga L, Brezzi F, Cangiani A, Manzini G, Marini LD, Russo A. Basic principles of virtual element methods. Mathematical Models and Methods in Applied Sciences. 2013; **23**(1):199-214. DOI: 10.1142/S0218202512500492

[17] Dragne C, Chiroiu V. Hexapod workspaces – Positions, speeds, forces. In: The 9th International Conference on Computational Mechanics and Virtual Engineering, Comec-2021, 21–23 October. Romania: Transilvania University of Brasov; 2021

[18] Ericson C. Real-time collision detection. In: The Morgan Kaufmann Series in Interactive 3-D Technology. Amsterdam: CRC Press; December 22, 2004. ISBN-10: 1558607323

[19] Haddadin S, De Luca A, Albu-Schäffer A. Robot collisions: A survey on detection, isolation, and identification.

IEEE Transactions on Robotics. 2017; **33**(6):1292-1312

[20] Dragne C, Chiroiu V. Advanced collision detection techniques. In: International Multi-Conference on Systems & Structures (SysStruc '19). Reșița. November 7-8, 2019. Available from: http://www.sysstruc19.uem.ro/sysstruc19.html

[21] Stewart D. A platform with six degrees of freedom. Proceedings of Institutions of Mechanical Engineering. 1965;**180**:371-386

[22] Merlet JP. Direct kinematics of parallel manipulators. IEEE Transactions on Robotics and Automation. 1993;**9**: 842-845

[23] Matsuishi M, Endo T. Fatigue of Metals Subjected to Varying Stress-Fatigue Lives under Random Loading. JSEM, Fukuoka, Japan: Proc. Kyushu District Meeting; 1968. pp. 37-40

[24] Hibbeler RC. Failure Theories. Mechanics of Materials. 8th ed. USA: Pearson Prentice Hall; 2011. pp. 520-522

[25] Dragne C, Chiroiu V, Iliescu M, Todirite I. Damage detection and smart warning for eventual structure failures in mechatronic systems. In: IEEE World Conference on Applied Intelligence and Computing AIC-2022. 2022. Available from: https://www.aic2022.scrs.in/

[26] https://www.mathworks.com/help, [Accessed: May, 2022]

[27] https://www.solidworks.com/, [Accessed: May, 2022]

[28] https://grabcad.com/ciprian.dragne, [Accessed: April, 2022]

[29] Paolo F. History of Robots and Robotic Surgery. The SAGES Atlas of Robotic: Surgery; 2018

[30] Bratovanov N, Zamanov V. Modeling and simulation of robots for semiconductor automation by using SolidWorks API. Proceedings of Technical University of Sofia. 2016; **66**(2):71-80

[31] Neto P, Pires JN, Moreira AP. Robot path simulation: A low cost solution based on CAD. In 2010 IEEE Conference on Robotics, Automation and Mechatronics. IEEE; June 2010. pp. 333-338

[32] Bratovanov N. Robot modeling, motion simulation and off-line programming based on SolidWorks API. In: Third IEEE International Conference on Robotic Computing (IRC). New York City, USA: IEEE Publisher; 2019. pp. 574-579. DOI: 10.1109/IRC.2019.00117

[33] Sosa-Méndez D, Lugo-González E, Arias-Montiel M, García-García R. ADAMS-MATLAB co-simulation for kinematics, dynamics, and control of the Stewart–Gough platform. International Journal of Advanced Robotic Systems. 2017;**14**:172988141771982. DOI: 10.1177/1729881417719824

Section 2

Robot Design and Control

Chapter 3

Biomechanical Design Principles Underpinning Anthropomorphic Manipulators

Mahonri William Owen and Chikit Au

Abstract

The biomechanical design of an artificial anthropomorphic manipulator is the focus of many researchers in diverse fields. Current electromechanical artificial hands are either in the research stage, expensive, have patents, lack severely in function, and/or are driven by robotic/mechanical principles, which tend to ignore the biological requirements of such designs. In response to the challenges addressed above this chapter discusses the potential of current technology and methods used in design to bridge the chasm that exists between robot manipulators and the human hand. This chapter elucidates artificial anthropomorphic manipulator design by outlining biomechanical concepts that contribute to the function, esthetics and performance of artificial manipulators. This chapter addresses joint stabilization, tendon structures and tendon excursion in artificial anthropomorphic manipulators.

Keywords: anthropomorphic, manipulator, biomechanics, design, mechanics, robotics

1. Introduction

In general, traditional mechanics and robotics have been the fundamental foundation for artificial hand/manipulator design. As a result, robotic and electro-mechanical manipulators lack the dexterity and anthropomorphism that comes from considering the physiological and biological aspects inherent in the human hand. This chapter looks at three key biomechanical inspired principles that directly influence artificial hand function and dexterity. These are:

1. Joint stabilization

2. Tendon structures

3. Tendon excursion

Figure 1 below gives an example of one potential anthropomorphic manipulator employing the above-mentioned principles.

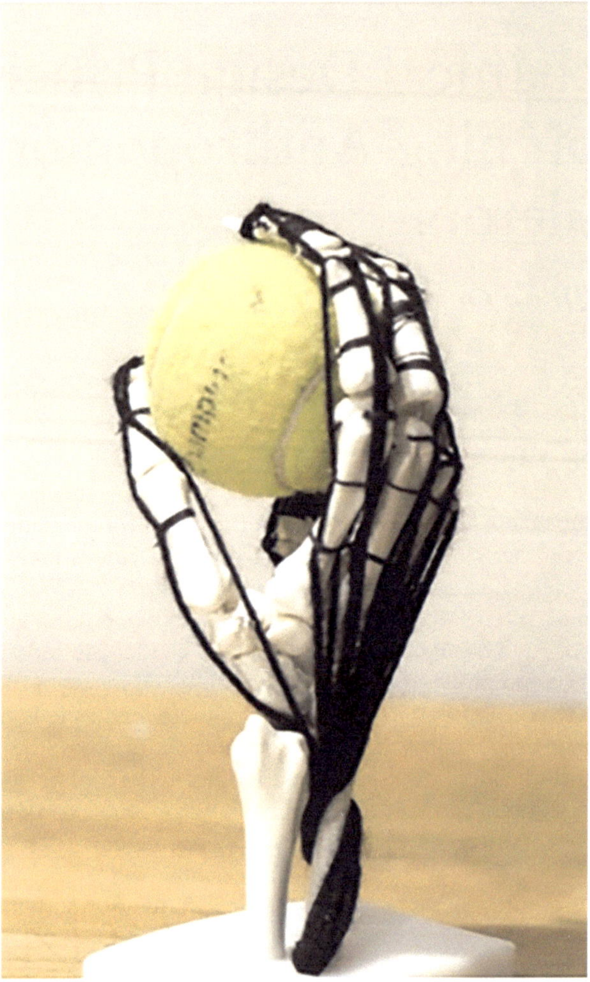

Figure 1.
Anthropomorphic approach toward artificial hand design.

1.1 Paradigm shift in anthropomorphic manipulator design

The very nature of human anatomy presents a unique challenge in the replication of the human hand. Mechanical engineering approaches rely on the expectation that the artificial hand design approach is quantifiable. The following reasons provided by [1] providing insight as to why mimicking the human body is not a straight forward process.

1. The boundaries of ligaments, tendons, and muscles are not easily definable.

2. Ligament, tendon, and muscle insertions differ from person to person.

3. Some tendons and ligaments possess non-linear characteristics that make measurement inaccurate.

4. Some human tissues have no measurable resting position.

For these reasons quantifiable methods are hard to realize. In reality, working in the space of anthropomorphic manipulator design becomes subject to engineer/designer best practice or previously knowledge. This reality often lends to this chapter, will express the value of a paradigm shift by exploring the biomechanical and physiological design principles that contribute to anthropomorphic manipulator design. The chapter concludes with a set of defined grasps that are a direct result of the design paradigm shift.

1.2 Current state

To this point in time, the human hand has been the "golden standard" of function, versatility and dexterity in grasping tasks as proven by the ongoing attempts to artificially recreate it [2]. Mimicking the qualities and properties of the human hand has always been a lofty and hard to realize goal, however, due to the current technological advancements of our age we are closer to achieve it than ever before. For these reasons this research is progressive, of great value and exciting. The ability of an artificial hand to function in an unmodified human environment directly relates to its capacity to function like the human hand. Traditional approaches to artificial hand design have been susceptible to ignore and/or omit the organic biological features of the human anatomy that are essential to hand function. Recent trends are tending toward artificial hand design that includes multidisciplinary viewpoints that consider more than just traditional mechanical principles.

In artificial hand design it is important to consider the Degrees of Freedom (DoF) the hand possesses. The DoF of a system is the number of independent parameters that define its configuration. The reviews [3, 4] describe some of the available electromechanical hands currently on the market [5–16]. Each review supports the general view that as the DoF of an artificial hand increases so does its anthropomorphism and function. Research into modeling the human hand claims that 24 DoF accurately represents the posture and movement of the human hand (including the wrist) [17]. There is consensus around this point with most literature agreeing that the human hand has between 21 and 26 DoF depending on whether you include or exclude the wrist [18]. The difference in DoF between studies can also be attributed to the DoF associated with the carpal bones of the hand and whether or not they are included.

Latter attempts at hand design employ the idea of bio-mimicry or an approach that is human inspired. The focus of these types of hands are on the tendinous structures actuating the hand and their synergies with ligaments, joints and actuators [4] approaches artificial hand design by mimicking the human bones and joints of the hand. Naturally, this makes for an esthetically pleasing design. In contrast to this approach a bulky unnatural looking wrist was implemented in the design and was recognized as required improvement [19] builds upon the idea of biomimicry by developing artificial tendinous structures that mimic the human hand and by improving the design of artificial thumbs. Their approach was esthetically pleasing and functioned well. Within their approach there was a new idea to incorporate joint capsules at each joint. Further studies [1] in the same year presented work on mimicking the mechanics and material properties possessed by the human hand and incorporating them into working artificial prototypes. The work presented herein emphasis the synergistic tendon networks, bone configuration, bone orientation, and joint development that are required in artificial hand design. It is expected that these types of approaches will increase in popularity and improve as the research climate and time allows.

2. Joint stabilization

The following section addresses the connection that exists between hand function and artificial ligament design. Firstly, it is important to understand how the bones of the human hand lay a foundation for the ligaments to layer on. The bones on which hand ligaments are inserted provide ligament pathways on which joint stabilization relies. Incorrect placement of ligaments severely effects the way a bone moves, transfers force and thus can drastically reduce the function of a hand. The bones of the human hand are essential to hand function, however, without the stabilizing ligaments and tendons they are of little use. Each joint of the hand has a series of ligaments that limit undesirable movement and protect the hand against excessive force. Hence, joint stabilization by virtue of the biological layers present in the human hand are integral to hand function.

Joint stabilization is integral to hand function; therefore, it is important to unpack the role and location of the ligaments responsible for finger joint stabilization and movement. In particular, the following section describes and defines the collateral ligament, the volar plate, and the annular ligament. Each ligament is then put in context by applying them to the design shown in **Figure 1**. We begin by explaining the collateral ligament and follow that by exploring both the volar plate and annular ligament.

2.1 Collateral ligament

This section describes the artificial collateral ligament by defining suitable insertion locations of the ligament for each phalanx of the hand that optimizes the movement between adjacent bones. The collateral ligament acts as the primary stabilizer of phalanx joints and is responsible for connecting bones. Taking the index finger as an example we name and orient the bones of the finger in **Figure 2**.

Determining the insertion positions of the collateral ligament on the artificial proximal phalanx (**Figure 3**) is defined by two features. These features are: The phalanx length, L_P, and the centroid, c. the phalanx length, L_P is the summation of L_{BC} and L_{HC} while the centroid is found through the CAD software SOLIDWORKS using the mass property evaluation.

The coordinate system is established on the centroid of the phalanx. The phalanx length and the location of the centroid determine the ligament insertion locations I_H and I_B. L_B and L_H define the limits that bound the insertion locations. The coordination of the head insertion I_H and base insertion I_B are given by

$$I_H = (x, y) = ((L_{HC} - L_{HI}), (y < 0)) \tag{1}$$

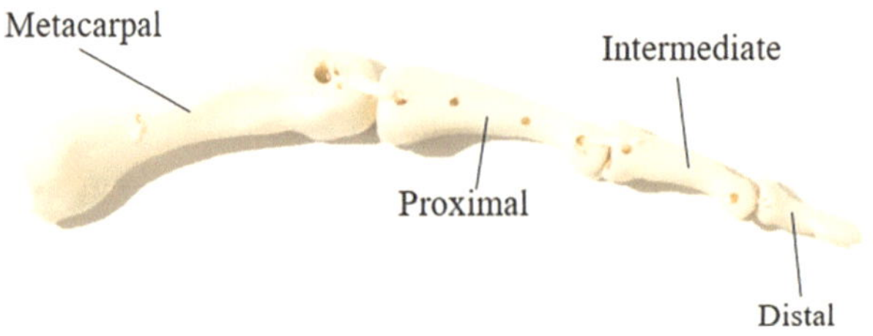

Figure 2.
Index finger bone configuration and orientation.

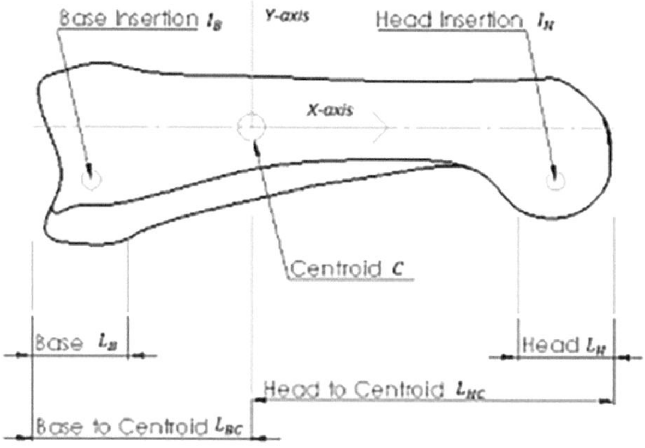

Figure 3.
Collateral ligament insertion points.

and

$$I_B = (x, y) = ((-L_{BC} + L_{BI}), (y < 0)) \tag{2}$$

Where L_{BI} and L_{HI} are determined by a ratio of the total phalanx length, L_P, that is still acceptable for correct bone interaction between each phalanx. The values for L_{BI} and L_{HI} are determined by a joint motion test. Insertion points are designed and modeled at distances between 3 mm and 8 mm from each end of the involved phalanx. Joint motion is quantified by a rating between one and three. Where one represents a good range of motion and three represents a poor or severely lacking range of motion. Any insertions below 3 mm or above 8 mm do not provide any range of motion and do not add contribute to the motion of the joint. (Values for L_{BI} and L_{HI} are found by dividing the insertion distance by the phalanx length). As shown in Eqs. (1) and (2) the location of the insertion in the vertical direction is governed by the variable, y. If, $y = 0$ the insertion is located along the x-axis. If $y < 0$ the insertion is located below the y-axis. If the insertion is governed by $(\pm y)$ then it has two insertion points either side of the y-axis. The three insertion location conditions are expressed by the following equations:

Condition (1) is expressed by

$$(x, y) = ((L_{HC} - L_{HI}), (y = 0)) \tag{3}$$

Condition (2) is expressed by

$$(x, y) = ((L_{HC} - L_{HI}), (\pm y)) \tag{4}$$

Condition (3) is expressed by

$$(x, y) = ((-L_{BC} + L_{BI}), (y < 0)) \tag{5}$$

For ease of application **Table 1** below outlines the conditions applicable to both the head and base of each phalanx.

Any major variance from the values and conditions listed in this table are detrimental to the joints range of motion and its stabilization. Through this method

	Metacarpal		Proximal		Intermediate		Distal	
	Head	**Base**	**Head**	**Base**	**Head**	**Base**	**Head**	**Base**
Condition 1			x		x			
Condition 2	x							
Condition 3				x		x		x

Table 1.
Insertion location conditions assigned to each phalanx.

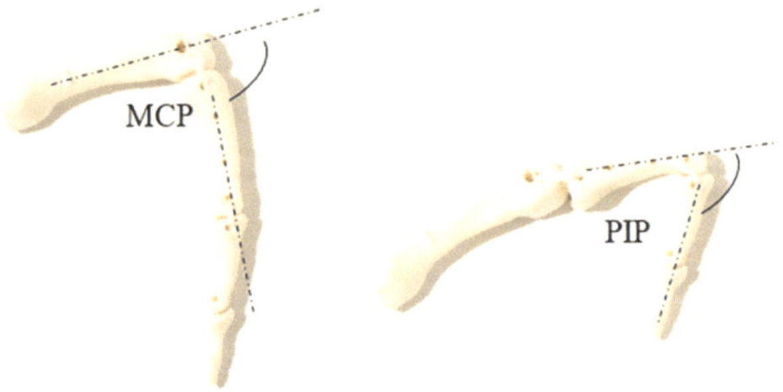

Figure 4.
MCP and PIP joints including their ranges of motion.

collateral ligament insertion is generalized for all artificial hand design. The geometry of the bones in the hand compliment the collateral ligaments at each bone-ligament interface. Collateral ligament placement determines joint function and therefore hand function. Bone surfaces are congruent and possess depressions or protuberances that give clues to ligament insertion locations. The bone surfaces also permit human like range of motion at each joint. **Figure 4** shows the resulting range of motion, μ for the metacarpophalangeal (MCP) and proximal interphalangeal (PIP) joints.

The resulting range of motion of the artificial MCP and PIP joints sit within two degrees measurement of their natural human counterparts, meaning digit function and dexterity are maximized. The collateral ligaments are vitally important to hand function, however, as recognized at the beginning of this section there are three main ligaments responsible for joint stabilization. The next section will look at the remaining ligaments which are the volar plate and the annular ligament.

2.2 Digit joints

The collateral ligament represents one of the three stabilizing ligaments of the joints of the fingers. This section will describe the role and placement of the volar plate and the annular ligaments with respect to the previously described artificial collateral ligament. The combination of these three ligaments make up the joints of the finger and work in harmony to provide the function and dexterity of the hand. The collateral ligament provides lateral stability and is represented in the black box **Figure 5**. The insertions are achieved via an extruded cut through the proximal phalanx and abide by the conditions stated in **Table 1**.

In synergy with the collateral ligament is the volar plate. The volar plate functions to protect the fingers against hyperextension with its insertion points highlighted in the red boxes of **Figure 5**. These insertions are found on the palmar side of the

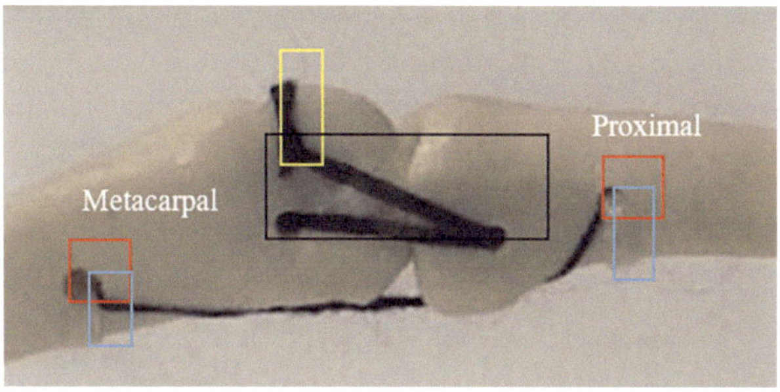

Figure 5.
The stabilizing ligaments of the MCP joint.

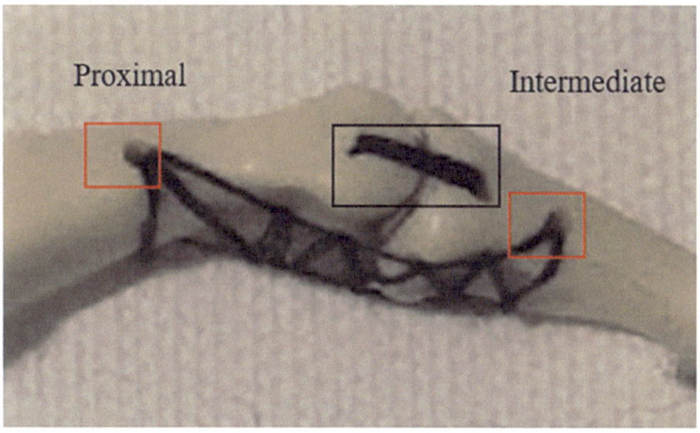

Figure 6.
The stabilizing ligaments of the PIP joint.

metacarpal head and the palmar side proximal base. A properly function volar plate limits rotation about the joint and absorbs excessive force placed on the joint in unwanted directions. It can most easily be thought of as a limiter of unwanted movement.

An artificial sagittal band is incorporated into the design of the artificial hand (yellow box of **Figure 5**). Although the sagittal band and the collateral ligament share insertion locations their functions are very different. The sagittal band creates a pathway for the extensor tendon responsible for force transferral and finger extension. It also eliminates transverse tendon slippage. In addition to the sagittal band the hand incorporates annular ligaments. These ligaments share insertions with the volar plate. These ligaments are light in color but outlined by blue boxes in **Figure 5**. The combination of these ligaments allows abduction/adduction and flexion/extension at the MCP joint. The lateral movement in the MCP joint is limited through the placement and angle of the collateral ligaments while rotation is limited by the volar plate. If designed correctly the congruence provided by the geometry of the bones allows smooth movement through the flexion and extension movements of the joint. The concave/convex relationship between the bones also allows abduction and adduction.

The PIP joint (**Figure 6**) connects the intermediate and proximal phalanges. The PIP joint on its own is a planar manipulator capable of only flexion and extension and just like the MCP is stabilized by the collateral ligament (black box **Figure 6**).

The palmar side of the PIP joint houses the volar ligament. The insertion points of this ligament are outlined in **Figure 6** by red boxes. This ligament functions to protect the joint from hyperextension and shares an insertion with the annular pulleys. The PIP joint provides simple planar movement that allows the hand to grasp and manipulate objects with precision and force. Bone geometry and size naturally limits the unwanted motion inherent in the joint, allowing a natural moving and anthropomorphic esthetic.

The final joint of the finger is the distal interphalangeal (DIP) joint. This joint connects the intermediate and distal phalanges. The DIP joint is a copy of the PIP joint described above.

2.3 The carpometacarpal joint

The thumb has traditionally been an incredibly hard appendage to mimic and replicate artificially. On its own the thumb is the single largest contributor to hand function and dexterity. It is what sets humans apart from the rest of the animal kingdom. With these sentiments in mind it is important to consider the biomechanical makeup of the thumb and incorporate it within the design process.

The Carpometacarpal (CMC) joint of the thumb provides the hand with complex multi-axial movement. The reason for this is the complex geometry of the trapezium bone and its relationship with the metacarpal of the thumb. Artificial ligaments provide stabilization for the CMC joint of the thumb on the artificial hand. The five artificial ligaments are based on the human counterparts and are described below.

The Anterior Oblique ligament (**Figure** 7) is the first of the stabilizing ligaments of the thumb. T1 represents the trapezium bone while M1 indicates the metacarpal bone of the thumb. The anterior oblique ligament functions to limit unwanted movement that would cause the CMC dislocation. The ligament connects the metacarpal bone and trapezium bones. The insertion points are highlighted in red.

In addition to the anterior oblique ligament there is the second stabilizing ligament shown in **Figure 8**. This ligaments function is to stabilize the joint and to limit the movement between the thumb and the palm. As shown, this ligament is inserted near

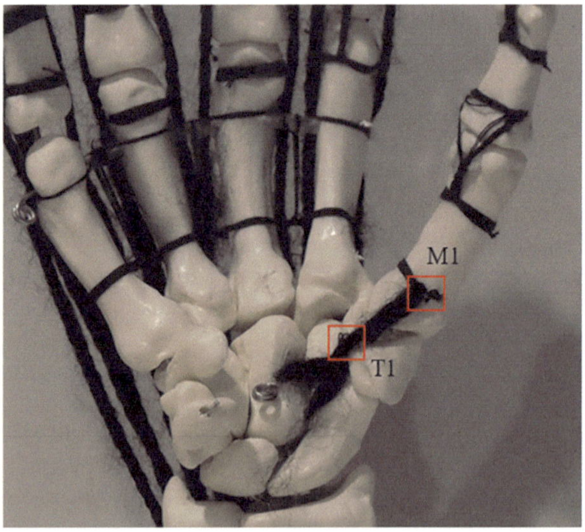

Figure 7.
The anterior oblique ligament of the CMC joint. (palmar view).

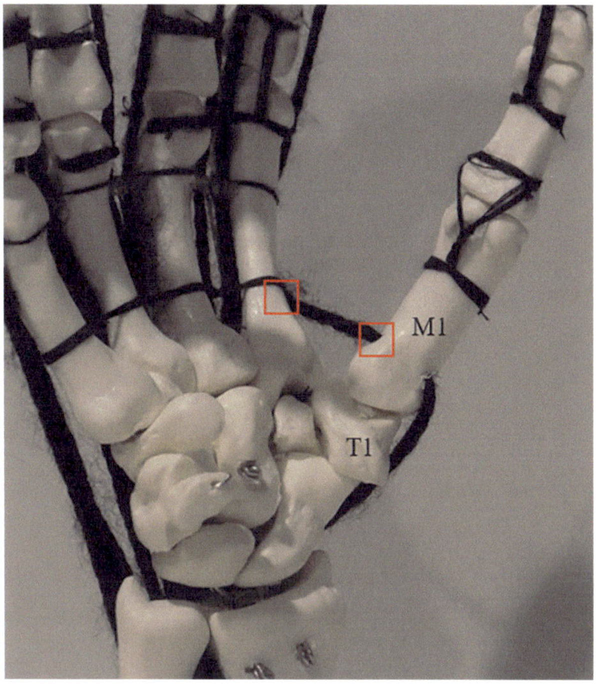

Figure 8.
CMC joint and its artificial ligaments (dorsal view).

the base of the index metacarpal and the thumb metacarpal (insertions are highlighted in red).

Ligaments three, four and five are artificial replications of the dorsal deltoid shaped ligaments. These are also stabilizing ligaments and are named the dorsal radial ligament, the dorsal central ligament and the posterior oblique ligament. These ligaments share an insertion on the trapezium bone and are connected the surface of the metacarpal bone of the thumb. Each ligament inserts at different points along the metacarpal surface to provide the thumb with a considerable range of motion. All three ligaments contribute to allowing and restricting of movement within their respective three-dimensional spaces. These three ligaments are shown **Figure 9** and are highlighted in red.

The previously mentioned ligaments work in synergy to stabilize the thumb joint and create the base on which the CMC joint relies for function. Along with the complex geometry of the bones at the CMC join these ligaments are responsible for the complex multi-axial movement of the thumb.

This and previous sections have conveyed one possible approach to joint stabilization in artificial anthropomorphic manipulators. The following section will now move into the actuation of these joints by tendons.

3. Tendinous structures

Movement of the human skeleton originates from the transferral of forces between it and its associated tendon/muscle. This is especially important for understanding hand movement as each minute movement can be attributed to the actuation of a

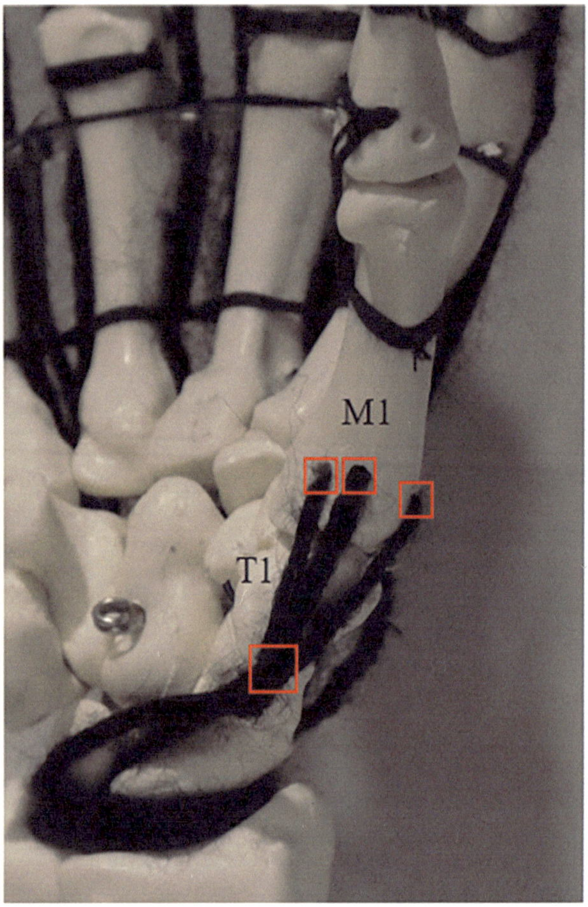

Figure 9.
The CMC joint and its artificial dorsal deltoid shaped ligaments (palmar/lateral view).

muscle or tendon. This section approaches the biomimicry of this process by separating the complex movement of the hand into two. Firstly, the tendons that control the fingers and secondly, the tendons that control the thumb.

3.1 Finger tendons

The finger has two primary movements and two minor movements. The primary movements are flexion and extension. The extensor tendon straightens the finger while the flexor tendons bend the finger. The minor movements of the finger are called adduction and abduction. These movements are induced by the contraction and elongation of the palmar interossei muscles. In order to understand the movements of these tendons and how they transfer force into the bones we need to look at their insertion points. We start by outlining the insertions of the extensor tendon. The first insertion is on the distal phalanx and the second is on the intermediate phalanx. These positions are highlighted in the red boxes in **Figure 10**. After its insertion on the intermediate phalanx the extensor tendon splits. This split helps to engage the finger in abduction and adduction. Artificial sagittal bands control the extensor tendon pathway as shown in **Figure 10** (white boxes) and prevent transverse slipping of the

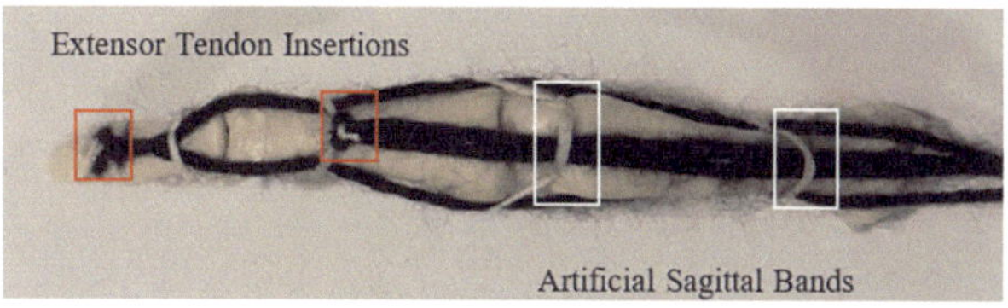

Figure 10.
The pathway and insertion points of the extensor tendon. (Dorsal view).

tendon. The flexion of the fingers in the hand are actuated by the Flexor Digitorum Profundus (FDP) tendon.

These FDP tendons traverse the palmar side of the hand and insert at the distal phalanx of each finger. The following section will outline the tendons of the thumb responsible for the thumb function, movement and anthropomorphism.

3.2 Thumb tendons

The thumb is a masterpiece of mechanical complexity and is capable of producing complex multi-axial movement. It is essential to understand its basic anatomy and function. Four main tendons are responsible for this movement.

- The Abductor Pollicis Longus (APL).
- The Flexor Pollicis Longus (FPL).
- The Extensor Pollicis Brevis (EPB).
- The Extensor Pollicis Longus (EPL).

The design and function of these tendons are described in the next section.

The artificial FPL shown in **Figure 11** (left-hand side) is channeled through the carpal tunnel and directed toward the base of the thumbs distal phalanx where it is inserted. The APL also shown in **Figure 11** (right-hand side) inserts into the thumbs metacarpal on the radial side and provides the thumb with abduction at the CMC joint.

Figure 12 (red boxes) below shows the insertion points of the EPB and the EPL. The EPB inserts into the thumbs proximal phalanx. This tendon is responsible for

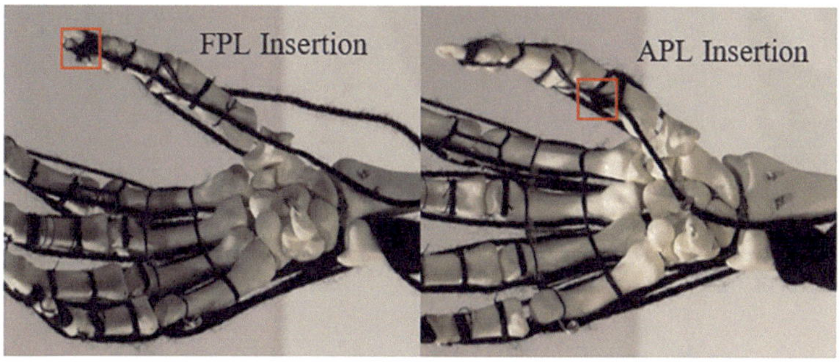

Figure 11.
The artificial FPL and APL insertions of the thumb. (Palmar view).

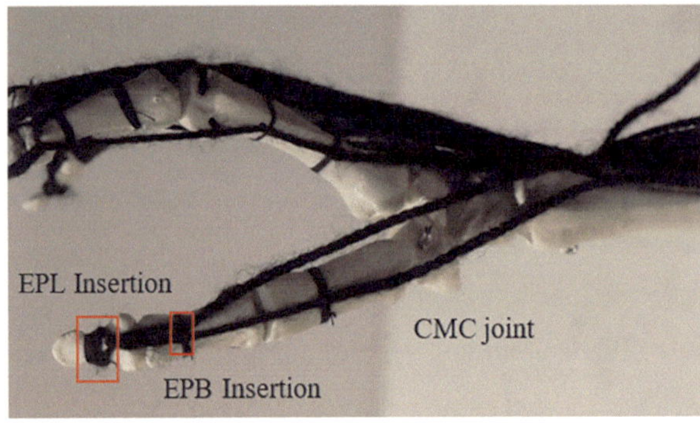

Figure 12.
The artificial EPL and EPB insertions of the thumb.

extension and abduction of the thumb at the CMC joint. The EPL is channeled through the rear of thumb and inserts at the tip distal phalanx. The EPL allows complete thumb extension whereas the PB can only provide partial extension.

It is through a complex synergy that the above-mentioned tendons provide the complex multi axial movement including flexion, extension, and circumduction.

4. Tendon excursion

Tendon excursion is often thought of as a limiting factor in electromechanical hand design. Tendon excursion is the displacement an artificial tendon experiences when the associated muscle/actuator contracts and induces tensile forces on it. Tendon excursion is can effect hand movement negatively.

For example, wrist movement induces tendon excursion on all the FDP tendons of the fingers. This can be seen in the natural flexion of the fingers during wrist extension and their natural extension during wrist flexion (**Figure 13**).

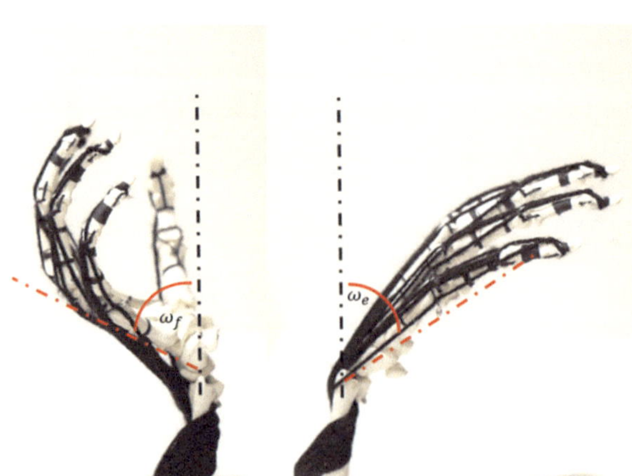

Figure 13.
Tendon excursion in the FDP of each finger induced by wrist movement.

The angles ω_f and ω_e represent the magnitude with which the wrist joint is in flexion or extension respectively. $\omega = 0$ is located vertically upward from the forearm bones and is measured from the pivot point of the wrist to the metacarpal creating the largest angle from the equilibrium point. In this case measurements are taken from the metacarpal of the little finger. Let us take the artificial FDP flexor tendon of each digit as an example. We can write a set of tendon displacements with respect to the wrist angle. Let, ω_f represent wrist flexion for any $\omega > 0$. We can write

$$\omega_f = \{d_{Fi} d_{Fm}\, d_{Fr}\, d_{Fl}\, d_{Ft}\} \tag{6}$$

Where, d is the displacement of the tendon in mm and the subscripts i, m, r, l and t are the index, middle, ring, little, and thumb digits respectively. The subscript, F, represents the flexor tendon and in the case of the thumb, $d_{Ft} = \{\, d_{Ft1}\, d_{Ft2}\}$. Where, d_{Ft1} *and* d_{Ft2} are artificial replications of the the APL and FPL (**Figure 11**) tendons respectively.

The effect of tendon excursion can be quantified by measuring the elongation/displacement flexor tendons while the wrist is moving. After measuring the artificial FDP displacement each finger was displaced by over 10 mm. A 10 mm displacement represents more than 50 percent of the fingers total motion, thus, tendon excursion cannot be ignored. In a similar manner the effect of tendon excursion on the thumb is too large to ignore, therefore we look to the following section as one solution to manage the unwanted effects of tendon excursion.

The source of displacement for each artificial tendon stems from its actuator, therefore, solutions can be found by adjusting the tendons actuator to account for the unwanted tendon excursion. Let us assume an anthropomorphic manipulator we have a series of four servomotors actuating four artificial FDP tendons. A servomotor correction factor can be implemented within the control scheme which considers the wrist angle. The correction factor allows the wrist induced excursion to meet equilibrium with the individual finger servomotors during wrist movement using:

$$e = \left(\frac{\pi R}{180}\right) \times \varphi_i \tag{7}$$

Where, R is the servo horn radius and φ_i is the angle of the servo horn. By so doing, the servomotor angles for each digit can be plotted against the wrist angle, ω at any given time.

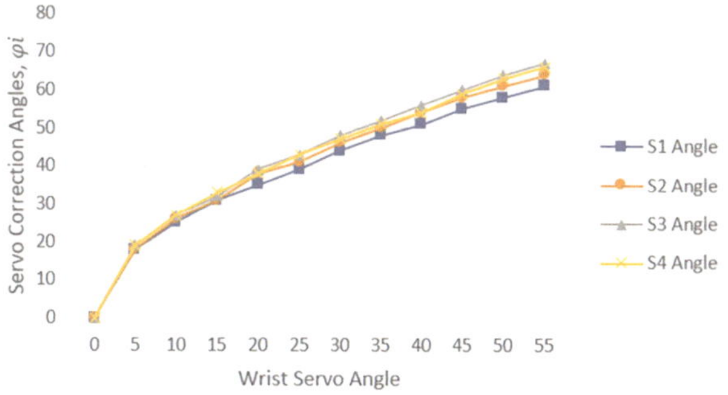

Figure 14.
Servomotor correction angles for wrist movement induced tendon excursion in the fingers.

The excursion correction angles displayed in **Figure 14** shows how servomotor angles change with respect to wrist position.

Including these types of correction tables into the control schemes of all electromechanical anthropomorphic hands is trivial but important for hand function and control.

5. Defining grasps

Artificial tendons provide movement in the hand. These tendons induce rotations about the joints of the digits. Assignments are shown in **Table 2**.

Figure 15 shows the joint angles of each phalanx with respect to each previous joint. Using the relationship between tendon displacement and joint angles we can define a grasp type.

Digit	Tendon label *i*	Digit movement
Thumb	1 and 6	Tendon 1 is for flexion and extension
Index finger	2	Tendon 6 is for opposition and reposition
Middle finger	3	Flexion and extension
Ring finger	4	Flexion and extension
Little finger	5	Flexion and extension

Table 2.
Tendon arrangement for finger actuation.

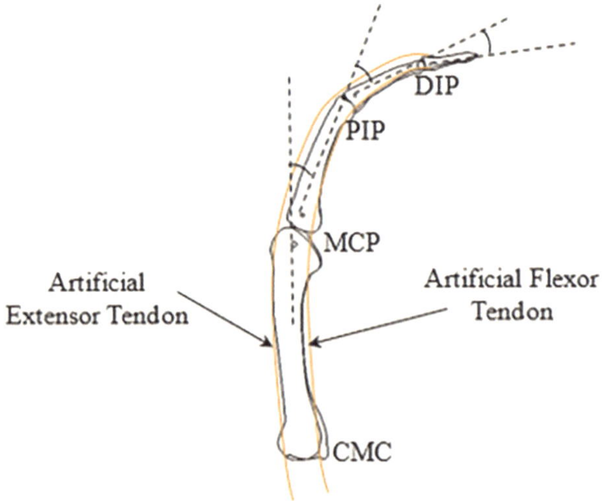

Figure 15.
Artificial hand finger tendon network and the associated joint angles.

For simple tendon networks composed of 3 pulleys in a digit, the tendon extension, e_i, is a linear function of the change of the joint angles $\Delta\theta_{i1}, \Delta\theta_{i2} and \Delta\theta_{i3}$. Therefore, tendon extension can be expressed as

$$e_i = \sum_j r_{ij}\Delta\theta_{ij} \tag{8}$$

Where, $\boldsymbol{r_{ij}}$, is a measure of the pulley radius at the *j*-th joint (*j* = 1, 2, 3 for *i* where $\boldsymbol{i = 1, 2, \dots, 5}$ and *j* = 1 for *i* = 6).

The extensions shown above can be used to determine grasp strength. Each tendon tendons is connected to individual servomotors to actuate digit movement. The artificial hand has six servomotors, one for each finger and two for the thumb. A grasp type in this context is defined as

$$\phi = [\varphi_1\varphi_2\varphi_3\varphi_4\varphi_5\varphi_6\ c]^T \tag{9}$$

Where,

$$c = \{1, for force closure grasping\ or\ 0, for form closure grasping$$

when φ_i (∀i) equals the pre-set maximum servomotor angle,φ_i^{max} the digit is in complete flexion. Otherwise, it is in complete extension (that is $\varphi_i = 0\ or \varphi_i^{min}$).

A physical representation of Eq. (9) is shown in the working principle of finger actuation displayed below in **Figure 16**.

In continuation of the rule established above **Figure 17** below demonstrates the four common grasp types and their definitions. The grasps shown are:

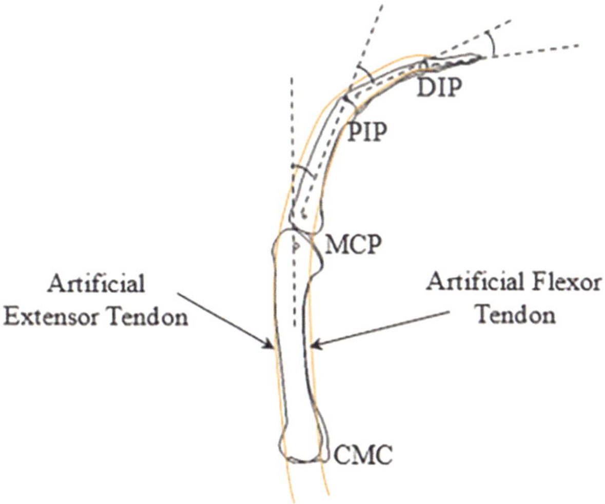

Figure 16.
Digit extension and flexion due to the servo motor angle.

Palmar pinch grasp

$\boldsymbol{\varphi} = [\varphi_1\ \varphi_2\ \varphi_3^{max}\ \varphi_4^{max}\ \varphi_5^{max}\ \varphi_6\ 1]^T$

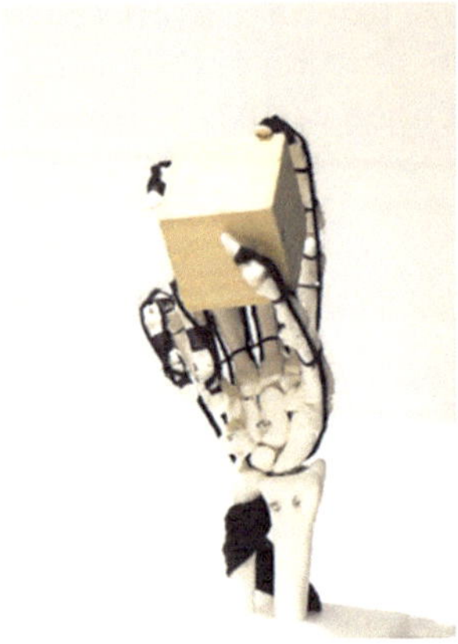

Tripod grasp

$\boldsymbol{\varphi} = [\varphi_1\ \varphi_2\ \varphi_3\ \varphi_4^{max}\ \varphi_5^{max}\ \varphi_6\ 1]^T$

Power grasp sphere (3 Finger) grasp

$\boldsymbol{\varphi} = [\varphi_1\ \varphi_2\ \varphi_3\ \varphi_4^{max}\ \varphi_5^{max}\ \varphi_6\ 0]^T$

Power sphere grasp

$\boldsymbol{\varphi} = [\varphi_1\ \varphi_2\ \varphi_3\ \varphi_4\ \varphi_5\ \varphi_6\ 0]^T$

Figure 17.
Grasp types performed by the artificial hand and their definitions.

- Palmar pinch.
- Tripod grasp.
- Power grasp sphere (3 finger).
- Power grasp.

In the case of objects where one of its dimensions are much larger than the other a tip pinch grasp or a power grasp is appropriate (**Figure 18**).

Tip pinch grasp
$\boldsymbol{\varphi} = [\varphi_1\ \varphi_2\ \varphi_3^{max}\ \varphi_4^{max}\ \varphi_5^{max}\ \varphi_6\ 1]^T$

Power grasp
$\boldsymbol{\varphi} = [\varphi_1\ \varphi_2\ \varphi_3\ \varphi_4\ \varphi_5\ \varphi_6\ 1]^T$

Figure 18.
Tip pinch and power grasp examples and definitions.

6. Conclusion

This chapter has outlined artificial anthropomorphic manipulator design by providing potential solutions to the challenges of mimicking the biomechanics of the human hand. Our current reality is that mechanical design considerations have been prioritized over bio-mimicry, this chapter approaches this challenge with a perspective that is refreshing and valuable. The biomechanical concepts explored in this chapter are vital in the design of artificial anthropomorphic manipulators. These biomechanical concepts contribute to the function, esthetics and performance of the manipulator and should not be ignored. It is expected that as joint stabilization, tendon structures and tendon excursion become prioritized that we will reach a level of anthropomorphism not realized in any other period in the history of artificial hand design.

Nomenclature, abbreviations and symbols

MCP	Metacarpophalangeal
PIP	Proximal Interphalangeal
DoF	Degrees of Freedom
DIP	Distal Interphalangeal
CMC	Carpometacarpal
FDP	Flexor Digitorum Profundus
APL	Abductor Policis Longus
FPL	Flexor Policis Longus
EPL	Extensor Policis Longus
EPB	Extensor Policis Brevis

L_p	Phalanx length
I_H	Head insertion
I_B	Base insertion
L_{HC}	Head to centroid length
L_{BC}	Base to centroid length
L_B	Base Length
L_H	Head Length
L_{BI}	Percentage of phalanx length
L_{HI}	Percentage of phalanx length
e_i	Tendon extension
ω_f	Wrist flexion
ω_e	Wrist extension
ω_f^{max}	Maximum wrist flexion
ω_e^{max}	Maximum wrist extension
φ_i	Servo horn angle
φ_i^{max}	Maximum servo horn angle
r_{ij}	radius of pulley
R	servo horn radius
T1	Trapezium bone
M1	metacarpal bone
e	tendon displacement
c	centroid/force-form closure
d	displacement
i	Index finger
m	Middle finger
r	Ring finger
l	Little finger
S1–S4	Servo motors

Author details

Mahonri William Owen* and Chikit Au
The University of Waikato, Hamilton, New Zealand

*Address all correspondence to: mowen@waikato.ac.nz

References

[1] Hockings N. Material and Mechanical Emulation of the Human Hand. England: University of Bath; 2016

[2] Belter JT, Segil JL, Dollar AM, Weir RF. Mechanical design and performance of anthropomorphic prosthetic hands: A review. Journal of Rehabilitation Research and Development. 2013;**50**(5):599-618

[3] Owen M. The Development of a Brain Controlled Robotic Prosthetic Hand. New Zealand: University of Waikato; 2015

[4] Deshpande AD, Xu Z, Vande Weghe MJ, Brown BH, Ko J, Chang LY, et al. Mechanisms of the anatomically correct testbed hand. IEEE/ASME Transactions on Mechatronics. 2011; **18**(1):238-250

[5] Touch Bionics Introduces App Controlled Hand. Gizmag. 2015. Available from: http://www. gizmag. com/i-limb-ultra-revolution/ 27150/

[6] RSLSTEEPER. Bebionic Features. 2015. Available from: http://bebionic.com/ the_hand/features

[7] Bridgwater LB et al. The robonaut 2 hand—Designed to do work with tools. Robotics and Automation. 2012; **1050–4792**:3425-3430

[8] Jacobsen SC, Wood JE, Knutti DF, Biggers KB. The UTAH/MIT dextrous hand: Work in progress. The International Journal of Robotics Research. 1984; **3**(4):21-50

[9] Nakano Y, Fujie M, Hosada Y. Hitachi's robot hand. Robotics Age. 1984; **6**(7):18-20

[10] De Laurentis KJ, Mavroidis C. Mechanical design of a shape memory alloy actuated prosthetic hand. Technology and Health Care. 2002;**10** (2):91-106

[11] Bekey GA, Tomovic R, Zeljkovic I. Control architecture for the belgrade/ USC hand. Dextrous Robot Hands. 1990; **1**:136-149

[12] Pellerin C. The salisbury hand. Industrial Robot: An international Journal. 1991;**18**(4):25-26

[13] Lin L, Huang H. NTU hand: A new design of dextrous hands. Transactions of the ASME. 1998;**120**:282-292

[14] Butterfab J, Grebenstein M, Liu H, Hirzinger G. DLR-hand 2: Next generation of a dextrous robot hand. International Conference on Robotics and Automation. 2001;**1**:109-114

[15] Bock O. Axon-bus prosthetic system with michaelangelo hand. Otto Bock Quality for Life. 2014;**2014**:1-14

[16] Yaoyao Hao MC, Cipriani C, Popovic DB, Chen W, Zheng X, Carozza MC. Controlling hand-assistive devices. IEEE Robotics and Automation. 2013;**40**:1070-9932

[17] Cobos S, Ferre M, Sanchez-Uran MA, Ortego J, Aracil R. Human hand descriptions and gesture recognition for object manipulation. Computer Methods in Biomechanics and Biomedical Engineering. 2010;**13**(3):305-317

[18] Gustus G, Stillfried G, Visser J, Jorntell H, Van Der Smagt P. Human hand modelling: Kinematics, dynamics, applications. Biological Cybernetics. 2012;**1**:1-15

[19] Xu Z, Todorov E. Design of a highly biomimetic anthropomorphic robotic hand towards artificial limb regeneration. In: Proceedings of the 2016 IEEE International Conference on Robotic and Automation. Stockholm Sweden; 2016. pp. 3485-3492. DOI: 10.1109/ICRA.2016.7487528

Chapter 4

An Intelligent Position-Tracking Controller for Constrained Robotic Manipulators Using Advanced Neural Networks

Dang Xuan Ba

Abstract

Nowadays, robots have become a key labor force in industrial manufacturing, exploring missions as well as high-tech service activities. Possessing intelligent robots for such the work is an understandable reason. Adoptions of neural networks for excellent control accuracies of robotic control systems that are restricted in physical constraints are practical challenges. This chapter presents an intelligent control method for position tracking control problems of robotic manipulators with output constraints. The constrained control objectives are transformed to be free variables. A simple yet effective driving control rule is then designed to force the new control objective to a vicinity around zeros. To suppress unexpected systematic dynamics for outstanding control performances, a new neural network is employed with a fast-learning law. A nonlinear disturbance observer is then used to estimate the neural estimation error to result in an asymptotic control outcome. Robustness of the closed loop system is guaranteed by the Lyapunov theory. Effectiveness and feasibility of the advanced control method are validated by comparative simulation.

Keywords: robotic manipulators, neural network, constrained control, motion control, simulations

1. Introduction

The world is now passing the Industry Revolution 4.0 in which robots have played a crucial role in industrial, manufacturing, discovering, rescuing and day-life activities. Excellent position controllers are required in most of industrial robots [1, 2]. However, in reality, it is not easy to achieve outstanding control precision with simple control structures due to unexpected influences of internal uncertain nonlinearities and unpredictable external disturbances in systematic dynamics [3–6]. Nevertheless, most real-life robot joints are restricted in certain physical ranges. Note that, few danger issues could be activated if the joints went over such the boundaries [4, 5]. To deal with the strict control problems, many research outcomes have been recently reported for both fully actuated and underactuated robotic systems [7, 8]. To realize control objective in predefined constraints, backstepping-based controllers are favorite approaches for developers [9, 10]. Barrier Lyapunov functions are employed as core-stones to implement the nonlinear control procedures [11, 12]. Such the advanced state-interfered techniques could cope with both static and dynamical practical constraints of robotic systems [11, 13]. As comparing to the backstepping-based methods, sliding-mode-control (SMC) approaches are also potential control solutions for output-constraint control problems thanks to the simpler design and implementation [14, 15]. Furthermore, the SMC ones could be upgraded with employment of soft boundaries to result in Prescribed-Performance Control (PPC) remedies which could maintain the control objectives within predefined control accuracies [16, 17].

To reach excellent control performances, the nonlinear behaviors of the robotic systems need to be compensated during the control process [14–18]. The uncertain functionalities could be modeled with classical approaches such as basic force/torque transformation or optimal-energy solutions or decomposition analyses [18, 19]. Such the classical methods seem to be effective with simple robotic systems since they highly depend on the system structure [7, 20]. To enhance the modeling performances, fast-estimation approaches were studied in the past few years based on time-delay estimation (TDE) technologies [21, 22]. lumped dynamics of the system are simply approximated from information of acceleration signals and input gain matrices selected [23, 24]. Owing to the simplicity in deployment, a vast of real-time applications have been developed using such the TDE algorithm [24, 25]. Since the acceleration signals are normally computed from the position signals using high-order time derivatives, measurement noises could be amplified reducing the estimation effect [26, 27]. In fact, to learn the systematic behaviors in a model-free manner, intelligent methods are also great solutions [28, 29]. Thanks to the ability of universal approximation, the system dynamics could be learnt under black-box models using Radial-basis function (RBF) networks [30–32] or Fuzzy-hybrid-networks [33–35]. Once the neural networks are integrated in the control process, the control error could be adopted as main excitation signals of the learning process.

Since the networks require abundant excitation signals to activate the learning processes, the intelligent controllers would sacrifice unexpected transient time to reach the excellent steady-state control outcomes [30, 36]. As a result, high learning rates were normally adopted in the classical learning rules to speed up the estimation processes. Note that, the conventional learn laws only ensure boundedness of the control errors instead of the learning errors [5, 37]. To create certain bounds of neural weighting coefficients, the networks were modified by integrating linear-leakage functions in their adaptation mechanisms [4–37]. Since the adaptation rules of free channels were not properly deactivated, the convergence processes of the overall systems are slower than those of the conventional ones. Note that although the nonlinear dynamics of the robotic systems could be efficiently compensated by the advanced neural networks, to yield outstanding transient control precision, the neural estimation errors need to be tackled [11, 13–38]. Integration of both neural networks and disturbance observers in nonlinear controllers have been proven to be excellent solutions for modern robotic systems [39–41]. Indeed, such the nonlinear integration was shown promising control results in stiffness-control robots, and in an exoskeleton, or in cooperative robots [42–44]. However, intelligent controllers using linear leakage functions in the neural adaptation rules require large robust signals to attain asymptotic control errors. With the combination of the neural network and disturbance in the intelligent approaches, the transient performances were remarkably improved, but the control errors were not still driven to zero in a smooth manner.

This chapter presents a new intelligent high-performance motion controller for robotic manipulators with output constraints. To deal with such the constraint problem, the control objective is first converted to a free variable using a new nonlinear transformation function. The indirect control objective is next driven to a certain vicinity using a sliding-mode-like control signal. A nonlinear neural network and disturbance observer are combined in a special fashion to construct a new closed-loop system in which both the estimation and control errors are pushed to zero in infinite time. The proposed controller possesses the following contributions:

- A novel nonlinear controller is proposed to stabilize the control objective inside arbitrary vicinity of zero without violation of the physical constraints.

- A nonlinear learning law of the neural network is developed to effectively estimate uncertain nonlinearities in the system model.

- To result in the asymptotic control performance of the overall system the neural estimation error is finally tackled by a nonlinear disturbance observer integrated.

- Working performances of the proposed control method is rigorously analyzed by an integral Lyapunov approach and extensive simulation results.

Outline of the paper is structured as follows. Section 2 presents the modeling of the studied systems and problem statements. Section 3 shows the design procedure of the proposed control algorithm with new neural disturbance estimation techniques and the stability analysis. Section 4 discusses the validation results obtained from comparative simulations. The conclusions are finally drawn in Section 5.

2. System model and problem statements

Behaviors of a general *nDOF* robot can be expressed using the following dynamics [19, 20]:

$$\mathbf{M}\lfloor\mathbf{q}\rfloor\ddot{\mathbf{q}} + \mathbf{C}\lfloor\mathbf{q},\dot{\mathbf{q}}\rfloor\dot{\mathbf{q}} + \mathbf{g}\lfloor\mathbf{q}\rfloor + \mathbf{f}\lfloor\dot{\mathbf{q}}\rfloor + \boldsymbol{\tau}_{\mathbf{d}} = \boldsymbol{\tau} \tag{1}$$

where $\boldsymbol{\tau} \in \Re^n$ is the vector of the control torques generated by joint actuators, $\mathbf{q} \in \Re^n$ is the vector of joint position or the system output, $\mathbf{M}\lfloor\mathbf{q}\rfloor \in \Re^{n\times n}$ is the positive-definite mass matrix, $\mathbf{C}\lfloor\mathbf{q},\dot{\mathbf{q}}\rfloor\dot{\mathbf{q}}, \mathbf{g}\lfloor\mathbf{q}\rfloor, \mathbf{f}\lfloor\dot{\mathbf{q}}\rfloor, \boldsymbol{\tau}_{\mathbf{d}} \in \Re^n$ are the Centripetal/ Coriolis vector, the gravitational, frictional, and the external disturbance torques, respectively.

Remark 1: The main control objective here is to derive a proper control signal $(\boldsymbol{\tau})$ to drive the system output $(\mathbf{q})$ following a desired trajectory $(\mathbf{q}_d)$.

Before designing the expected control approach, the following assumptions are consolidated.

Assumption 1 [45, 46]: The disturbance $(\boldsymbol{\tau}_{\mathbf{d}})$ is bounded and Lipschitz continuous.

Assumption 2: the reference profile $(\mathbf{q}_{\mathbf{d}})$ is known, bounded and twice continuously differentiable.

Assumption 3: The system states $(\mathbf{q}, \dot{\mathbf{q}})$ are measurable.

Remark 2: The robotic system Eq. (1) is a passivity model with bounded time-derivative states [19, 22, 45]. For practical systems, the robot joints $(\mathbf{q})$ are limited in physical ranges:

$$\underline{\mathbf{q}} \leq \mathbf{q} \leq \overline{\mathbf{q}} \tag{2}$$

where $\left(\underline{\mathbf{q}} \text{ and } \overline{\mathbf{q}}\right)$ are respectively the lower and upper bounds of the system output $(\mathbf{q})$.

In reality, unexpected impacts from physical collisions could make the system danger.

Remark 3: To obtain an excellent controller for the stated problem, one needs a proper control strategy that could deal with dynamical nonlinear behaviors of the robotic system Eq. (1) in complying with the physical constraint and be able to drive the control objective to zero as fast as possible. Furthermore, the controller is also expected to be robust and model-free.

3. Intelligent nonlinear constrained controller

A robust adaptive controller is designed in this section based on a new constrained sliding mode framework and new learning mechanism of a basic neural network and nonlinear disturbance observer. Stability of the closed control system is then investigated by Lyapunov theories.

3.1 Constrained sliding mode control with neural network

We first define the following control error as the main control objective:

$$\mathbf{e} = \mathbf{q} - \mathbf{q_d} \tag{3}$$

The error is in fact allowed to vary in the following range that is constructed by the constraint (2).

$$\begin{cases} \underline{\mathbf{e}} \leq \mathbf{e} \leq \overline{\mathbf{e}} \\ \overline{\mathbf{e}} \equiv \overline{\mathbf{q}} - \mathbf{q_d} > 0 \\ \underline{\mathbf{e}} \equiv \underline{\mathbf{q}} - \mathbf{q_d} < 0 \end{cases} \tag{4}$$

where $(\underline{\mathbf{e}} \text{ and } \overline{\mathbf{e}})$ are the lower and upper physical bounds of the control error$(\mathbf{e})$, respectively.

The following transformation function is next proposed to map the constrained error $(\mathbf{e})$ to a new free space:

$$y_{i|i=1..n} = \frac{e_i}{(\overline{e}_i - e_i)(e_i - \underline{e}_i)} \tag{5}$$

where $\mathbf{y} = [y_1, y_2, \ldots, y_n]^T$ is the transformed error, e_i is a specific entry of the control error vector $\mathbf{e} = [e_1, e_2, ..e_i., e_n]^T$.

A sliding manifold is defined as an indirect control objective of the studied system:

$$\mathbf{s} = \dot{\mathbf{e}} + \mathbf{K}_0\mathbf{y} \tag{6}$$

where $\mathbf{K}_0 = \text{diag}\lfloor \mathbf{k}_0 \rfloor = \text{diag}\lfloor [k_{01}; \ldots; k_{0n}] \rfloor$ is a positive-definite diagonal gain matrix.

The time derivative of the manifold Eq. (6) under the dynamics Eq. (1) is expressed

$$\dot{\mathbf{s}} = -\mathbf{v} + \overline{\mathbf{M}}^{-1}\boldsymbol{\tau} - \ddot{\mathbf{q}}_\mathbf{d} + \mathbf{K}_0\dot{\mathbf{y}} \tag{7}$$

where, $\mathbf{v} = -\mathbf{M}^{-1}(\mathbf{C}\dot{\mathbf{q}} + \mathbf{g} + \mathbf{f} + \boldsymbol{\tau}_\mathbf{d}) + \left(\mathbf{M}^{-1} - \overline{\mathbf{M}}^{-1}\right)\boldsymbol{\tau} \in \Re^n$ is defined as a systematic-deviation term that is composited from both the internal dynamics and external disturbances, and $\overline{\mathbf{M}} = \text{diag}\lfloor [\overline{m}_1, \ \overline{m}_2, \ \ldots, \ \overline{m}_m] \rfloor$ is a nominal positive-definite mass matrix selected.

Based on the manifold system Eq. (7), the final control signal is structured from a dynamical control term ($\boldsymbol{\tau}_{\mathbf{DYN}}$), error-driving term ($\boldsymbol{\tau}_{\mathbf{DRI}}$), and robust control term ($\boldsymbol{\tau}_{\mathbf{ROB}}$), as follows:

$$\boldsymbol{\tau} = \overline{\mathbf{M}}(\boldsymbol{\tau}_{\mathbf{DYN}} + \boldsymbol{\tau}_{\mathbf{DRI}} + \boldsymbol{\tau}_{\mathbf{ROB}}) \tag{8}$$

The dynamical signal ($\boldsymbol{\tau}_{\mathbf{DYN}}$) is used to compensate for the internal dynamics ($\mathbf{v}$) in Eq. (7). With robotic manipulators, the lumped dynamics ($\mathbf{v}$) are bounded [19, 42] but very complicated and not easy to derive [20]. To study such the complex behaviors, a neural network could be thought of a reasonable tool. The dynamics $\mathbf{v} = [v_1, v_2, \ \ldots, v_n]^T$ can be modeled using the following universal linear combination:

$$v_{i|i=1..n} = \mathbf{w}_i^T \mathbf{r}_i \lfloor \mathbf{q}, \ \dot{\mathbf{q}}, \ \boldsymbol{\tau} \rfloor + \delta_i \tag{9}$$

where $\mathbf{w}_i$,$\mathbf{r}_i\lfloor \mathbf{q}, \ \dot{\mathbf{q}}, \ \boldsymbol{\tau} \rfloor$,$\delta_i$ are optimal weight vectors, neural regression vectors, and the modeling error, respectively.

Hence, the signal is structured as follows:

$$\boldsymbol{\tau}_{\mathbf{MOD}} = \hat{\mathbf{v}} + \ddot{\mathbf{q}}_\mathbf{d} - \mathbf{K}_0\dot{\mathbf{y}} \tag{10}$$

where, the approximation $\hat{v}_i$ is estimate of the dynamics v_i, and is designed as [17, 42]

$$\hat{v}_{i|i=1..n} = \hat{\mathbf{w}}_i^T \mathbf{r}_i \lfloor \mathbf{q}, \ \dot{\mathbf{q}}, \ \boldsymbol{\tau} \rfloor \tag{11}$$

in which, $\hat{\mathbf{w}}_i$ is estimate of the weight vector $\mathbf{w}_i$.

By employing the dynamical control signal Eq. (10), the dynamics Eq. (7) become

$$\dot{\mathbf{s}} = \tilde{\mathbf{v}} + \boldsymbol{\tau}_{\mathbf{DRI}} + \boldsymbol{\tau}_{\mathbf{ROB}} \tag{12}$$

where $\tilde{\mathbf{v}} = [\tilde{v}_1, \tilde{v}_2, \ \ldots, \tilde{v}_n]^T = \hat{\mathbf{v}} - \mathbf{v} \in \Re^n$ is an estimation-error vector.

Since the role of the control signal ($\boldsymbol{\tau}_{\mathbf{DRI}}$) is to drive the sliding manifold to around zero from an arbitrary initial position, it is selected as

$$\boldsymbol{\tau}_{\mathbf{DRI}} = -\mathbf{K}_1\mathbf{s} \tag{13}$$

where $\mathbf{K}_1 = \text{diag}\lfloor \mathbf{k}_1 \rfloor = \text{diag}\lfloor [k_{11}; \ldots; k_{1n}] \rfloor$ is a diagonal positive-definite gain matrix.

Since the robust control signal $\boldsymbol{\tau}_{\mathbf{ROB}}$ is adopted to suppress the estimation error $(\boldsymbol{\delta})$, it is designed as

$$\boldsymbol{\tau}_{\mathbf{ROB}} = -\mathbf{K}_2 \operatorname{sgn} \lfloor \mathbf{s} \rfloor \tag{14}$$

where $\mathbf{K}_2 = \operatorname{diag}\lfloor \mathbf{k}_2 \rfloor = \operatorname{diag}\lfloor [k_{21}; \ldots; k_{2n}] \rfloor$ is a diagonal positive-definite gain matrix.

The manifold dynamics Eq. (12) is now expressed as

$$\dot{\mathbf{s}} = \tilde{\mathbf{v}} - \mathbf{K}_1 \mathbf{s} - \mathbf{K}_2 \operatorname{sgn} \lfloor \mathbf{s} \rfloor \tag{15}$$

Remark 3: The dynamics Eq. (15) indicate that the closed-loop system is bounded stable if the estimate $(\hat{\mathbf{v}})$ is bounded. In theoretical aspects, the asymptotic control performance would be resulted in if the robust gain $(\mathbf{k}_2)$ is selected satisfying a condition of $(\mathbf{k}_2 > \boldsymbol{\delta})$. However, with such a big robust control gain, it could activate chattering phenomena. In contrast, a small robust gain could reduce the control precision.

To approximate the nonlinear dynamics $(\mathbf{v})$, the network Eq. (11) is activated using the control information of the sliding manifold under following rule:

$$\dot{\hat{\mathbf{w}}}_i = -\operatorname{diag}\lfloor \mathbf{a}_{1i} \rfloor \operatorname{diag}\lfloor \mathbf{r}_i^2 \rfloor \frac{s_i^2}{1 + s_i^2} \hat{\mathbf{w}}_i - b_i s_i \mathbf{r}_i \tag{16}$$

where $\mathbf{B} = \operatorname{diag}\lfloor [b_1,\ b_2,\ \ldots,\ b_n] \rfloor$ and $\operatorname{diag}\lfloor \mathbf{a}_{1i} \rfloor_{i=1..n}$ are diagonal positive-definite constant matrices.

The control performance of the neural-constrained sliding mode system is investigated by the following statements.

Theorem 1: By employing the robust control rule Eqs. (3)–(14) and the neural learning law Eqs. (11), Eq. (16) to control the robotic system Eq. (1) under the output constraint Eq. (2), the closed-loop system is asymptotically stable if the control gains comply with

$$\begin{cases} k_{1i|i=1..n} > \dfrac{1}{4\varepsilon_i} \mathbf{r}_i^T \operatorname{diag}\lfloor \mathbf{r}_i \rfloor \operatorname{diag}\lfloor \mathbf{a}_{1i} \rfloor \left| \mathbf{w}_i^2 \right|^T \\ k_{2i} > |\delta_i|_{\max} \end{cases} \tag{17}$$

The proof of Theorem 1 is given in Appendix A.

Remark 4: As seen in Eq. (15), once the real dynamics $(\mathbf{v})$ are well estimated with an arbitrary small accuracy, small robust gains would yield good control performances. Note that, approximation by the neural network is a multi-channel learning work. The learning rule Eq. (16) is hence designed to increase the neural updating effect.

4. Disturbance-observer integration

The neural-constrained nonlinear control structure provides the excellent control performances with stationary trajectory signals. In high-speed working frequencies, the estimation error $(\boldsymbol{\delta})$ becomes large and could degrade the control accuracy. Adoption of an additional control term based on the disturbance-observer technology could be an understandable solution. The following assumption could be taken into account:

Assumption 4: The error ($\boldsymbol{\delta}$) and its time derivative are bounded. It could be thus modeled as a first-order system:

$$\dot{\boldsymbol{\delta}} = -\text{diag}\lfloor\boldsymbol{\alpha}\rfloor\boldsymbol{\delta} + \boldsymbol{\zeta} \tag{18}$$

where $\text{diag}\lfloor\boldsymbol{\alpha}\rfloor = \text{diag}\lfloor[\alpha_i,\ \alpha_2,\ \dots,\ \alpha_n]\rfloor$ is a diagonal positive-definite constant matrix.$\boldsymbol{\zeta} = [\varsigma_1, \varsigma_2, \dots, \varsigma_n]^T$ is a virtual bounded disturbance vector.

To effectively compensate for the estimation error ($\boldsymbol{\delta}$), the robust control signal Eq. (14) is updated with a disturbance-estimation term, as follows:

$$\boldsymbol{\tau}_{\mathbf{ROB}} = -\mathbf{K}_2\,\text{sgn}\,\lfloor\mathbf{s}\rfloor + \hat{\boldsymbol{\delta}} \tag{19}$$

at which $\hat{\boldsymbol{\delta}} = \left[\hat{\delta}_1,\ \hat{\delta}_2,\ \dots,\ \hat{\delta}_n\right]^T$ is estimate vector of the neural modeling error $(\boldsymbol{\delta})$. It is computed from the following learning rule:

$$\dot{\hat{\boldsymbol{\delta}}} = -\text{diag}\lfloor\boldsymbol{\alpha}\rfloor\hat{\boldsymbol{\delta}} - \mathbf{BP}^{-1}\mathbf{s} - \mathbf{K}_3\,\text{sgn}\,\lfloor\mathbf{s}\rfloor \tag{20}$$

Here, $\mathbf{P} = \text{diag}\lfloor[p_1,\ p_2,\ \dots,\ p_n]\rfloor$,$\mathbf{K}_3 = \text{diag}\lfloor[k_{31},\ k_{32},\ \dots,\ k_{3n}]\rfloor$ are diagonal positive-definite constant matrices.

Validation results in previous work [46, 47] confirmed the learning efficiency of the disturbance observer Eq. (20) for simple systems. To connect the disturbance observer with the neural sliding mode control scheme, the adaptation rule of the network is improved as

$$\begin{aligned}\dot{\hat{\mathbf{w}}}_{i|i=1..n} = &-\text{diag}\lfloor\mathbf{a}_{1i}\rfloor\text{diag}\lfloor\mathbf{r}_i^2\rfloor\frac{s_i^2}{1+s_i^2}\hat{\mathbf{w}}_i - \text{diag}\lfloor\mathbf{a}_{2i}\rfloor\text{diag}\lfloor\mathbf{r}_i^2\rfloor\hat{\mathbf{w}}_i\\ &-\left(b_i s_i + p_i k_{3i}\,\text{sgn}\,\lfloor s_i\rfloor\right)\mathbf{r}_i\end{aligned} \tag{21}$$

where $\text{diag}\lfloor\mathbf{a}_{2i}\rfloor_{i=1..n}$ is a diagonal positive-definite constant matrix.

The stability of the closed-loop system is validated by the following statement.

Theorem 2: By employing the robust control rule Eqs. (3)–(14) combining with the neural learning law Eqs. (11), Eq. (21) and disturbance observer Eqs. (19), (20) to control the robotic system Eq. (1) under the output constraint Eq. (2), the closed-loop system is asymptotically stable if the control gains comply with

$$\begin{cases} k_{1i|i=1..n} > \dfrac{1}{4b_i}\mathbf{r}_i^T\text{diag}\lfloor\mathbf{r}_i\rfloor\text{diag}\lfloor\mathbf{a}_{1i}\rfloor\left|\mathbf{w}_i^2\right|^T \\ k_{2i} > 0 \\ k_{3i} > |\varsigma_i|_{\max} + \dfrac{1}{4p_i k_{2i}}\mathbf{r}_i^T\text{diag}\lfloor\mathbf{r}_i\rfloor\text{diag}\lfloor\mathbf{a}_{2i}\rfloor\left|\mathbf{w}_i^2\right|^T \end{cases} \tag{22}$$

The proof of Theorem 2 is discussed in Appendix B.

Remark 5: From Theorem 2, it can be seen that the robust control gain $(\mathbf{k}_2)$ could be selected with a small value for a high control accuracy. Obviously, the robustness of the closed-loop system is undertaken by a large value of the disturbance-observer gain$(\mathbf{k}_3)$.Remark 6: After the sliding manifold $(\mathbf{s})$ converges to zero, the control error $(\mathbf{e})$ will approach to origin under the sliding phase [25, 27]. Adoption of the nonlinear synthetization (9) could speed up the convergence time of the sliding process [17, 18]. The detailed block diagram of the proposed controller is presented in **Figure 1**.

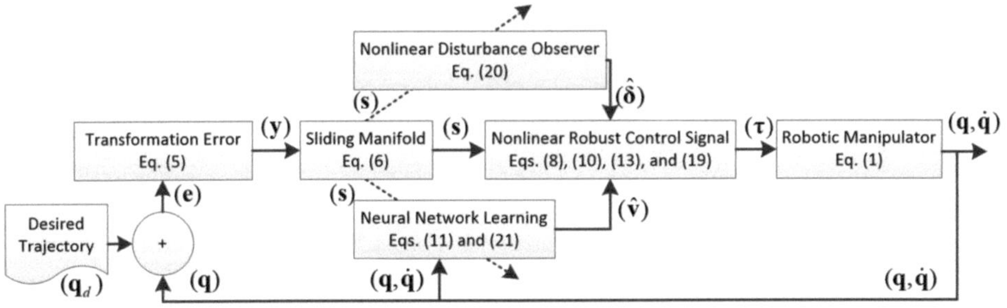

Figure 1.
Structure of the proposed controller.

5. Verification results

Validation results of the developed controller in various testing conditions are discussed are discussed in this section. To provide the competitive evaluation, a classical Proportional-Integral-Derivative (PID) controller and linear neural-disturbance-observer (LND) controller were also implemented to control the same system in the same working conditions. The LND algorithm was referred from previous and is re-expressed in Appendix C.

The controllers were employed for motion control of a 3DOF robot, as depicted in **Figure 2**. Detailed dynamics of the 3DOF robot were derived based on the Lagrange method [4, 19, 47], as formulated in Appendix D. The neural network had 9 inputs $(q_i, \dot{q}_i, \tau_i)_{|i=1,2,3}$ and 730 neurons with the logsig activation

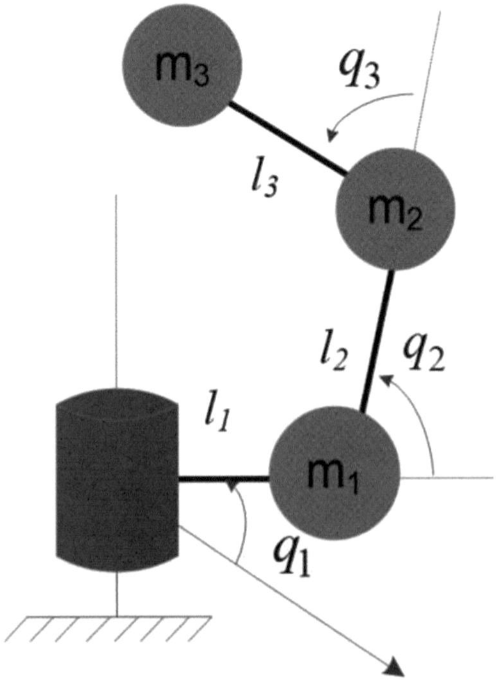

Figure 2.
Configuration of the simulation 3DOF robot.

Description	Parameters	Values	Unit
Link length 1	l_1, l_2, l_3	0.1, 0.2, 0.2	m
Gravitational Accel	g	9.81	m/s^2
Friction coefficient	a_1, a_2, a_3	20, 20, 20	N.s
Mass of links	m_1, m_2, m_3	5, 3, 2	kg

Table 1.
Detailed parameters of the simulation model.

Description	Parameters	Values
	LND Controller [45]	
Nominal mass matrix	$\overline{\mathbf{M}}$	$\mathbf{I}_3$
Control gains	$\mathbf{K}_{c0}, \mathbf{K}_{c1}$	diag([10; 100; 10]), diag([200; 100; 10])
Disturbance gain	$\mathbf{K}_{c3}$	$30\mathbf{I}_3$
Learning rate	$\Gamma_{i\|i=1..3}$	$500\mathbf{I}_3$
Learning rates	$\mu_{i\|i=1..3}$	$0.002\mathbf{I}_3$
	PID	
Control gains	K_P, K_I, K_D	diag([700; 900; 500]), diag([50; 10; 10]), diag([10; 10; 10]),
	Proposed Controller	
Nominal mass matrix	$\overline{\mathbf{M}}$	$\mathbf{I}_3$
Control gains	$\mathbf{K}_0, \mathbf{K}_1, \mathbf{K}_2$, and $\mathbf{K}_3$	diag([100; 100; 10]), diag([40; 100; 2]), $0.1\mathbf{I}_3, 200\mathbf{I}_3$.
Leakage rates	$\mathbf{a}_{1i\|i=1..3}, \mathbf{a}_{2i}$	$10\mathbf{I}_3, 10\mathbf{I}_3$
Excitation rates	$\mathbf{B}, \mathbf{P}$	$200\mathbf{I}_4, 10\mathbf{I}_4$
Disturbance gain	$\alpha_{i\|i=1..3}$	2

Table 2.
Selected parameters of the controllers.

function in the hidden layer [42, 45]. All of the initial values of the weight vectors $\left(\hat{\mathbf{w}}_{i|i=1,2,3}\right)$ were set to be zero. Other simulation parameters of the dynamics and the controllers are shown in **Tables 1** and **2**, respectively. The control results obtained by the controllers are intensively discussed in the following subsections.

6. Simulation results

In the first simulation, the desired profiles were sinusoidal signals with different frequencies (0.1 (Hz), 0.3 (Hz), and 0.5 (Hz)), as plotted in **Figure 3**. Physical ranges of the robot joints were set to be $[-0.5\pi;\ 0.5\pi]$. The control results obtained are compared in **Figures 4–6**.

As shown in **Figure 5**, stability of the closed-loop system could be maintained by the PID controller and with good control errors: 1 (deg), 1.39 (deg) and 5.6 (deg) for joints 1, 2, and 3, respectively. However, as carefully observed in the response of joint 2 in **Figure 4**, the physical constraints were violated by the PID control in the transient time. To void the unexpected collision and provide high control performances both in the transient and steady-state phases, a combination of neural network, disturbance-observer learning techniques and the constrained backstepping control signal was adopted in the LND controller. Indeed, outstanding control

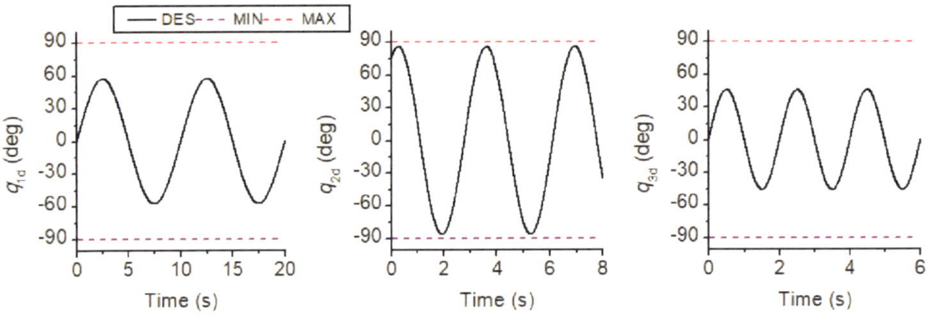

Figure 3.
The desired profiles of the robot joints in the first simulation.

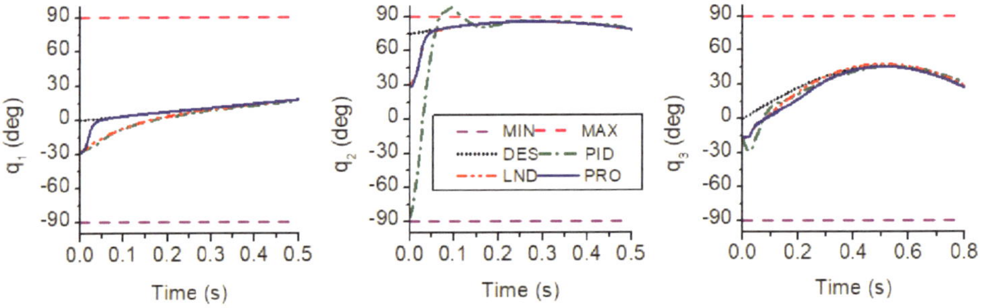

Figure 4.
System responses of the controllers obtained in the first simulation.

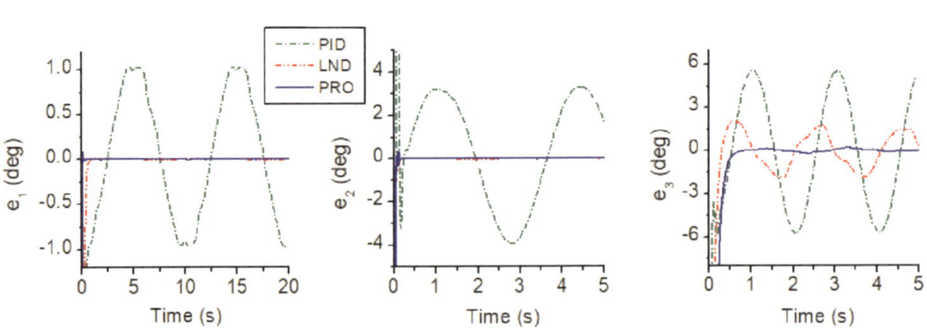

Figure 5.
Comparative control errors of the controller in the first simulation.

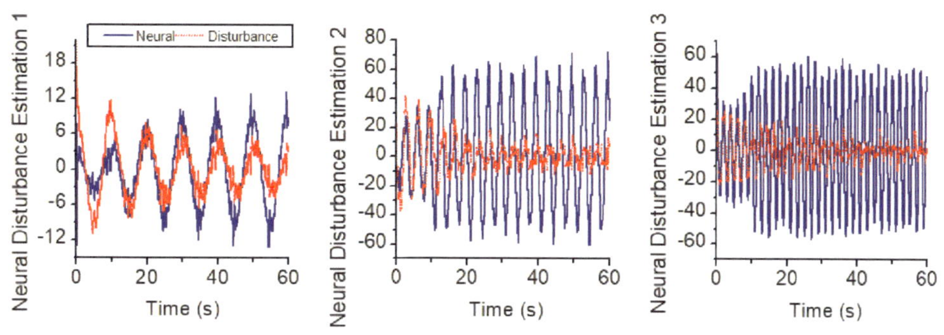

Figure 6.
Learning performances of the proposed controller in the first simulation.

precision was resulted in by the LND one: the control precision at joints 1,2, and 3 were 0.094 (deg), 0.105 (deg), and 1.85 (deg), respectively. The output-constraint control problem could be also dealt with by the constrained control algorithm designed. Moreover, the nonlinear dynamics of the robotic system were eliminated well by the proposed neural-disturbance learning method. As a result, higher control performances were delivered by the studied controller: the control accuracies at joints 1 and 2 were 0.11 (deg) and 0.108 (deg), respectively. The control results in **Figure 5** imply that although the control performances of the LND and proposed controllers were almost same in low-speed work conditions, they were clearly different in the high-frequency trajectory-tracking control. To this end, the proposed control algorithm was employed the new nonlinear learning rule Eq. (21) to improve the estimation effect, which are revealed from estimation data presented in **Figure 6**.

The controllers were continuously challenged with new various frequencies of the sinusoidal trajectories in the second test. The new frequencies at joints 1, 2, and 3 were selected to be 1 (Hz), 0.3 (Hz), and (0.7 Hz), respectively. **Figure 7** presents pieces of the new reference signals with respect to time. Applying the same controllers to the robotic system, the results obtained are shown in **Figure 8**.

Control results in **Figure 8** indicate that the PID control accuracies were seriously degraded in arduous testing conditions: the control errors were increased to 9.32 (deg) and 8.18 (deg) at joints 1 and 2, respectively. The LND control method could however maintain acceptable control performances thanks to the merging linear learning algorithm: control precision at joints 1 and 3 was slightly increased to be 0.7 (deg) and 4.1 (deg), respectively. Note that as discussed in previous work [45], it is difficult to

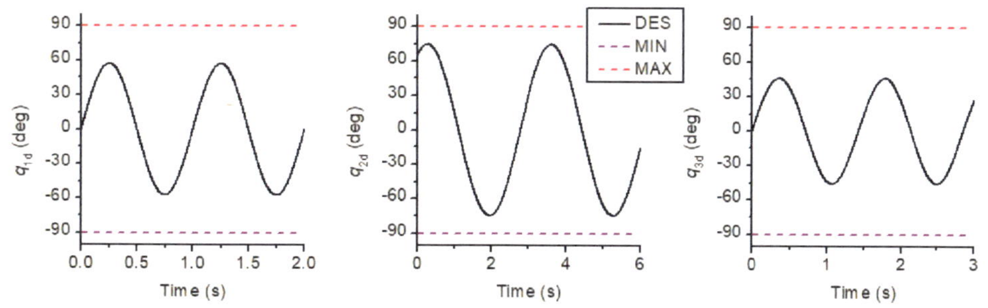

Figure 7.
The desired profiles of the robot joints in the second simulation.

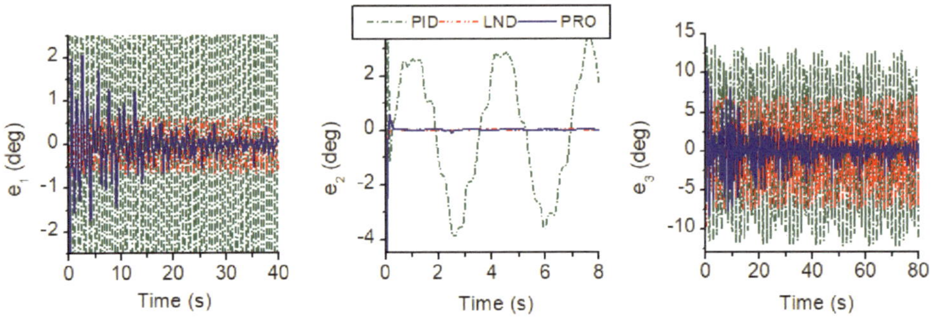

Figure 8.
Comparative control errors in the second simulation.

result in asymptotic control outcomes by the LND approach. This drawback was well overcome by the new learning rule proposed, in which the nonlinear network and disturbance observer were properly combined with an arbitrary small robust gain to ensure the asymptotic convergence of the closed-loop system. The convergences of the control errors obtained by the proposed control algorithm, as demonstrated in **Figure 8**, show that the uncertain nonlinearities and external disturbances in the system dynamics were well estimated by the collaborative nonlinear adaptation laws. The control and learning effectiveness of the new control approach was confirmed by the validation results achieved.

7. Additional discussion

By comparing the control results obtained by the two intelligent controllers, as presented in **Figures** 5 and **8**, it can be seen that their control performances would be same in the steady-state phases but really different in the transient phases. The nonlinear learning integration led to the faster learning effect and higher control precision.

Estimation data illustrated in **Figures** 7 and **9** imply that the neural network played as a crucial role in approximating the system dynamics, and the estimation error was then learnt by the nonlinear disturbance observer. Furthermore, with the merging control technique proposed, one only needs an arbitrary small robust signal to result in asymptotic control outcomes, that ensures the smooth control behaviors as presented in **Figure 10**.

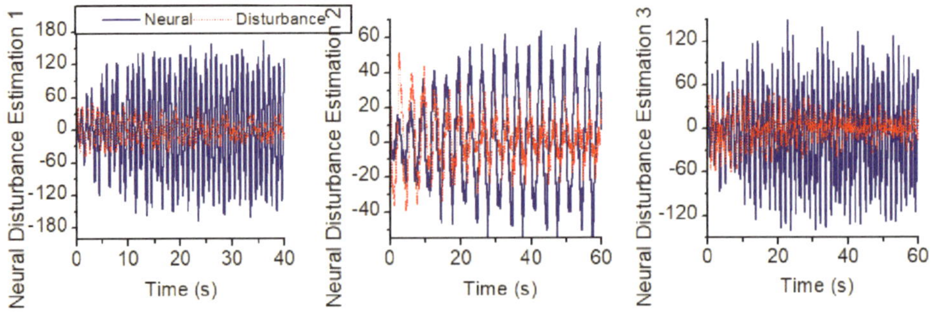

Figure 9.
Learning performances of the proposed controller in the second simulation.

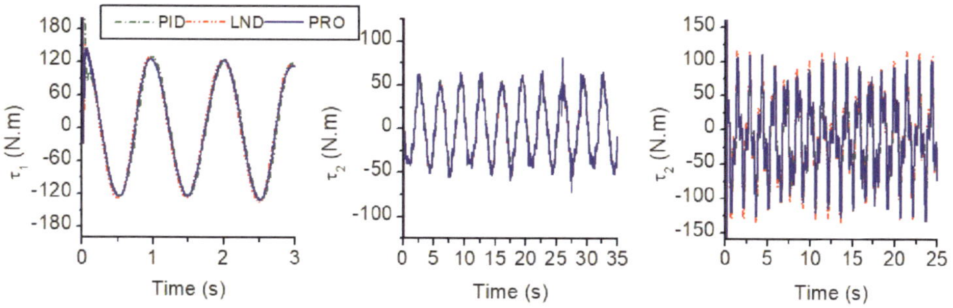

Figure 10.
Control signals generated by the controllers in the second simulation.

Control Error		Joint 1		Joint 2		Joint 3	
		MA	**RMS**	**MA**	**RMS**	**MA**	**RMS**
The 1st case	**PID**	1.02	0.69	3.9	2.5	5.8	4.1
	LND	0.096	0.051	0.12	0.044	1.94	1.21
	PRO	0.1.	0.008	0.11	0.025	0.41	0.041
The 2nd case	**PID**	9.3	6.3	3.9	2.3	8.2	5.5
	LND	0.7	0.44	0.13	0.02	4.1	2.6
	PRO	0.16	0.04	0.12	0.02	0.83	0.22

Table 3.
Statistical control errors of the comparative controllers.

Table 3 summarizes the maximum absolute (MA) and root-mean-square (RMS) values of the control errors for a specific working time (75 s to 85 s). As seen in the table, the best RMS errors were always provided by the designed controller even though its MA values were not the highest one in some cases. Here, we propose a ratio of RMS/MA values to deeply evaluate the control performances of the controllers in which those of the PID, LND and proposed controllers were in range of 0.64, 0.41, and 0.31, respectively. The smaller factors imply that the internal deviation and external disturbances were effectively eliminated by the corresponding controllers in better manners. The superior control performances of the proposed controller over the previous control methods are thus confirmed again by the intensive analyses based on the obtained results.

8. Conclusions

This chapter presents a new intelligent control method for high-performance motion control of robotic manipulators with output constraints. The controller is built based on a new neural-disturbance constrained sliding mode structure. A nonlinear sliding-mode control signal is first derived to strictly stabilize the control objective within predefined output constraints. The control accuracies are next improved by eliminating the nonlinear uncertainties and external disturbances in the system dynamics using a new nonlinear neural network. The estimation error is then compensated by proper integration of a nonlinear disturbance observer. By adoption of this neural-disturbance mechanism and a minor robust signal, an asymptotic control outcome is resulted in. The effectiveness of the overall control system is investigated by the rigorous theoretical proofs and comparative simulation results in various working conditions.

Appendix A: proof of theorem 1

The following Lyapunov function is first considered:

$$L_1 = 0.5\sum_{i=1}^{n}\left(b_i s_i^2 + \tilde{\mathbf{w}}_i^T \tilde{\mathbf{w}}_i\right) \tag{23}$$

By substituting the dynamics Eqs. (15) and (16) into the time derivative of the function (A1), we next have

$$\begin{aligned}\dot{L}_1 &= -\sum_{i=1}^{n}\left(b_i k_1 s_i^2 + |s_i|(k_2+\delta_i)\right) - \sum_{i=1}^{n}\left(\frac{s_i^2}{1+s_i^2}\tilde{\mathbf{w}}_i^T \operatorname{diag}\lfloor \mathbf{a}_{1i}\rfloor \operatorname{diag}\lfloor \mathbf{r}_i^2\rfloor \hat{\mathbf{w}}_i\right) \\ &\le -\sum_{i=1}^{n}\left(b_i k_1 s_i^2 + |s_i|(k_2+\delta_i)\right) - \sum_{i=1}^{n}\left(\frac{s_i^2}{1+s_i^2}\tilde{\mathbf{w}}_i^T \operatorname{diag}\lfloor \mathbf{a}_{1i}\rfloor \operatorname{diag}\lfloor \mathbf{r}_i^2\rfloor \tilde{\mathbf{w}}_i\right) \\ &+\sum_{i=1}^{n}\left(\frac{s_i^2}{1+s_i^2}|\tilde{\mathbf{w}}_i|^T \operatorname{diag}\lfloor \mathbf{a}_{1i}\rfloor \operatorname{diag}\lfloor \mathbf{r}_i^2\rfloor |\mathbf{w}_i|\right)\end{aligned} \tag{24}$$

From the condition Eq. (17), there always exist two positive constants $\lambda_{i1|i=1..n}, \lambda_{i2}$ for the following constraint:

$$\dot{L}_1 \le -\sum_{i=1}^{n}\left(\lambda_{i1} b_i k_1 s_i^2 + \lambda_{i2}\frac{s_i^2}{1+s_i^2}\tilde{\mathbf{w}}_i^T \operatorname{diag}\lfloor \mathbf{a}_{1i}\rfloor \operatorname{diag}\lfloor \mathbf{r}_i^2\rfloor \tilde{\mathbf{w}}_i\right) \tag{25}$$

It means that the proof of Theorem 1 has been completed.

Appendix B: proof of theorem 2

Dynamics of the subsystems Eqs. (15), (19), (20) in element-wise forms are presented as follows:

$$_{i=1..n}\begin{cases}\dot{s}_i = \tilde{\mathbf{w}}_i^T \mathbf{r}_i - k_1 s_i - k_{2i}\operatorname{sgn}\lfloor s_i\rfloor + \tilde{\delta}_i \\ \dot{\tilde{\delta}}_i = -\alpha_i\tilde{\delta}_i - \dfrac{b_i}{p_i}s_i - k_{3i}\operatorname{sgn}\lfloor s_i\rfloor - \varsigma_i\end{cases} \tag{26}$$

A new integral-type Lyapunov function is investigated:

$$L_2 = L_1 + \sum_{i=1}^{n}\left(0.5 p_i \tilde{\delta}_i^2 + \int_{s_i\lfloor 0\rfloor}^{s_i}(k_{3i}\operatorname{sgn}\lfloor s_i\rfloor + \varsigma_i)ds_i + L_{20i}\right) \tag{27}$$

where $L_{20i|i=1..n}$ is a positive constant selected as [46]:

$$L_{20i} = \frac{\left(k_{3i} + |\varsigma_i|_{\max}\right)^2}{2b_i} + \left(k_{3i} + |\varsigma_i|_{\max}\right)s_i\lfloor 0 \rfloor \tag{28}$$

The time derivative of the function (B2) under the dynamical behaviors Eqs. (B1) and (21) is constrained in the following inequality:

$$\begin{aligned}\dot{L}_2 \le & -\sum_{i=1}^{n}\left(b_i s_i (k_1 s_i + k_{2i}\,\text{sgn}\lfloor s_i \rfloor) + \alpha_i p_i \tilde{\delta}_i^2\right) \\ & -\sum_{i=1}^{n}\left(\tilde{\mathbf{w}}_i^T\left(\text{diag}\lfloor \mathbf{a}_{1i} \rfloor \frac{s_i^2}{1+s_i^2} + \text{diag}\lfloor \mathbf{a}_{2i} \rfloor\right)\text{diag}\lfloor \mathbf{r}_i^2 \rfloor \tilde{\mathbf{w}}_i\right) \\ & -\sum_{i=1}^{n}\left(p_i (k_1 s_i + k_{2i}\,\text{sgn}\lfloor s_i \rfloor)(k_{3i}\,\text{sgn}\lfloor s_i \rfloor + \varsigma_i)\right) \\ & +\sum_{i=1}^{n}\left(|\tilde{\mathbf{w}}_i|^T\left(\text{diag}\lfloor \mathbf{a}_{1i} \rfloor \frac{s_i^2}{1+s_i^2} + \text{diag}\lfloor \mathbf{a}_{2i} \rfloor\right)\text{diag}\lfloor \mathbf{r}_i^2 \rfloor |\mathbf{w}_i|\right)\end{aligned} \tag{29}$$

If the gains selected satisfying Eq. (22), there always exist another constant $\lambda_{i3|i=1..n}$ for the following inequality:

$$\begin{aligned}\dot{L}_1 \le & -\sum_{i=1}^{n}\left(\lambda_{i1} b_i k_1 s_i^2 + \alpha_i p_i \tilde{\delta}_i^2\right) \\ & -\sum_{i=1}^{n}\left(\tilde{\mathbf{w}}_i^T\left(\lambda_{i2}\frac{s_i^2}{1+s_i^2}\text{diag}\lfloor \mathbf{a}_{1i} \rfloor + \lambda_{i3}\text{diag}\lfloor \mathbf{a}_{2i} \rfloor\right)\text{diag}\lfloor \mathbf{r}_i^2 \rfloor \tilde{\mathbf{w}}_i\right)\end{aligned} \tag{30}$$

It leads to the proof of Theorem 2.

Appendix C: re-deign of a comparative linear neural-disturbance-observer backstepping controller

From a previous work [45], A linear neural-disturbance-observer backstepping (LND) controller is re-designed here for validation. Note that, the previous controller is developed in the single system space. From the control error Eq. (3), a virtual control signal $u_{i|i=1..n}$ and virtual control error $z_{i|i=1..n}$ are chosen as

$$_{i=1..n}\begin{cases} u_i = -k_{c0i}e + \dot{q}_{di} \\ z_i = \dot{q}_i - u_i \end{cases} \tag{31}$$

where $k_{c0i|i=1..n}$ are positive control gains.

The final control signal of the system is then selected as

$$\tau_{i|i=1..n} = -e_i - k_{c1i}z_i - \hat{\varphi}_i + u_i + \hat{\mathbf{w}}_i^T \boldsymbol{\psi}_i\lfloor \mathbf{q}, \dot{\mathbf{q}}, \dot{\mathbf{z}}, \dot{\mathbf{u}} \rfloor \tag{32}$$

where k_{c1i} are positive control gains, $\boldsymbol{\psi}_{i|i=1..n}$ are the regression vectors of the neural network. $\hat{\mathbf{w}}_{i|i=1..n}$ are estimates of the weight vector $\boldsymbol{\psi}_{i|i=1..n}$, and are updated by:

$$\dot{\hat{\mathbf{w}}}_i = -\Gamma_i(\boldsymbol{\psi}_i z_i + \mu_i \hat{\mathbf{w}}_i) \quad (33)$$

where μ_i are positive leakage rates, and Γ_i are diagonal positive-definite matrices.

$\hat{\varphi}_{i|i=1..n}$ are estimates of systematic disturbances, and are computed throughout an auxiliary variable $\hat{\phi}_{i|i=1..n}$, that is estimated by the following learning mechanism:

$$\begin{cases} \hat{\varphi}_i = \hat{\phi}_i + k_{c2i} z \\ \dot{\hat{\phi}}_i = -k_{c2i} \overline{m}_i^{-1} \left(\tau_i - \dot{q}_i + \hat{\varphi}_i\right) \end{cases} \quad (34)$$

where, k_{c2i} is a positive disturbance gain selected.

Appendix D: Dynamics of the simulation 3DOF robot

The dynamics (1) of the robot whose configuration is presented in **Figure 2**, can be derived in detail using the Euler–Lagrange method as follows:

$$\begin{cases} \mathbf{M}\lfloor \mathbf{q} \rfloor = \begin{bmatrix} m_{11} & 0 & 0 \\ 0 & m_{22} & m_{23} \\ 0 & m_{32} & m_{33} \end{bmatrix} \\ m_{11} = m_1 l_1^2 + m_2 \left(l_1 + l_2 \cos\left(q_2\right)\right)^2 \\ \quad + m_3 \left(l_1 + l_2 c_2 + l_3 \cos\left(q_2 + q_3\right)\right)^2 \\ m_{22} = m_2 l_2^2 + m_3 \left(l_2^2 + l_3^2 + 2 l_2 l_3 \cos\left(q_3\right)\right) \\ m_{23} = m_{32} = m_3 \left(l_2 l_3 \cos\left(q_3\right) + l_3^2\right) \\ m_{33} = m_3 l_3^2 \end{cases} \quad (35)$$

$$\begin{cases} \mathbf{C}\lfloor \mathbf{q},\ \dot{\mathbf{q}} \rfloor \dot{\mathbf{q}} = \left[c_1; c_2; c_3\right]^T \\ c_1 = -2m_2 \left(l_1 + l_2 \cos\left(q_2\right)\right) l_2 \sin\left(q_2\right) \dot{q}_2 \dot{q}_1 \\ \quad -2m_3 \left(l_1 + l_2 \cos\left(q_2\right) + l_3 \cos\left(q_2 + q_3\right)\right) \left(l_2 \dot{q}_2 \sin\left(q_2\right) + \left(\dot{q}_2 + \dot{q}_3\right) l_3 \sin\left(q_2 + q_3\right)\right) \dot{q}_1 \\ c_2 = -2m_3 l_2 l_3 s_3 \dot{q}_2 \dot{q}_3 - m_3 l_2 l_3 s_3 \dot{q}_3^2 + l_2 s_2 (l_1 + l_2 c_2) m_3 \dot{q}_1^2 \\ \quad - m_3 (-l_2 s_2 - l_3 s_{23})(l_1 + l_2 c_2 + l_3 c_{23}) \dot{q}_1^2 \\ c_3 = m_3 \left(l_2 l_3 \sin\left(q_3\right) \dot{q}_3\right) \dot{q}_2 + m_3 \left(l_1 + l_2 \cos\left(q_2\right) + l_3 \cos\left(q_2 + q_3\right)\right) l_3 \sin\left(q_2 + q_3\right) \dot{q}_1^2 \\ \quad + m_3 l_2 l_3 \sin\left(q_3\right) \dot{q}_2^2 + m_3 l_2 l_3 \sin\left(q_3\right) \dot{q}_2 \dot{q}_3 \end{cases} \quad (36)$$

$$\mathbf{g}\lfloor \mathbf{q} \rfloor = -g_0 \left[0; \left(2 l_2 \cos\left(q_2\right) + l_3 \cos\left(q_2 + q_3\right)\right); l_3 \cos\left(q_2 + q_3\right)\right]^T \quad (37)$$

$$\mathbf{f}\lfloor \dot{\mathbf{q}} \rfloor = \left[a_1 \dot{q}_1; a_2 \dot{q}_2; a_3 \dot{q}_3\right]^T \quad (38)$$

where q_i,l_i,m_i and $a_{i|i=1,2,3}$ are joint positions, link lengths, link masses and frictional coefficients, respectively; g_0 is the absolute gravitational-acceleration value.

Author details

Dang Xuan Ba
Ho Chi Minh City University of Technology and Education (HCMUTE),
Ho Chi Minh City, Vietnam

*Address all correspondence to: badx@hcmute.edu.vn

References

[1] Goel R, Gupta P. Robotics and industry 4.0. In: A Roadmap to Industry 4.0: Smart Production, Sharp Business and Sustainable Development. Cham: Springer; 2019. pp. 157-169

[2] Shehu N, Abba N. The role of automation and robotics in buildings for sustainable development. Journal of Multidisciplinary Engineering Science and Technology. 2019;**6**(2): 9557-9560

[3] Park Y, Jo I, Lee J, Bae J. A dual cable hand exoskeleton system for virtual reality. Mechatronics. 2018;**49**:177-186

[4] He W, Chen Y, Yin Z. Adaptive neural network control of an uncertain robot with full-state constraints. IEEE Transactions on Cybernetics. 2016; **46**(3):620-629

[5] Wang Y, Yan F, Chen J, Ju F, Chen B. A new adaptive time-delay control scheme for cable-driven manipulators. IEEE/ASME Transactions on Mechatronics. 2019;**15**(6):3469-3481

[6] Tri NM, Ba DX, Ahn KK. A gain-adaptive intelligent nonlinear control for an electrohydraulic rotary actuator. International Journal of Precision Engineering and Manufacturing, vo. 2018;**19**(5):665-673

[7] Guo Q, Liu Y, Wang Q, Jiang D. Adaptive neural network control of two-DOF robotic arm driven by electro-hydraulic actuator with output constraint. In: Proceedings of the IET Conference. Guiyang, China, 19–22, June; 2018

[8] Cui R, Guo J, Mao Z. Adaptive backstepping control of wheeled inverted pendulums models. Nonlin. Dyn. 2015;**79**(1):501-511

[9] Tee KP, Ren B, Ge SS. Control of nonlinear systems with time-varying output constraints. Automatica. 2011; **47**(11):2511-2516

[10] Cui R, Gao B, Guo J. Pareto-optimal coordination of multiple robots with safety guarantees. Autonomous Robots. 2012;**32**(3):189-205

[11] Wu L, Yan Q, Cai J. Neural network-based adaptive learning control for robot manipulators with arbitrary initial errors. IEEE Access. 2019;**7**: 180194-180204

[12] Zhou Q, Wang L, Wu C, Li H, Du H. Adaptive fuzzy control for nonstrict-feedback systems with input saturation and output constraint. IEEE Trans. Syst. Man Cybern. Syst. 2017;**47**:1-12

[13] Yu J, Zhao L, Yu H, Lin C. Barrier Lyapunov functions-based command filtered output feedback control for full-state constrained nonlinear systems. Automatica. 2019;**105**:71-79

[14] Zhu YK, Qiao JZ, Guo L. Adaptive sliding-mode disturbance observer-based composite control with prescribed performance of space manipulators for target capturing. IEEE Transactions on Industrial Electronics. 2019;**66**(3): 1973-1983

[15] Jing C, Xu H, Niu X. Adaptive sliding-mode disturbance rejection control with prescribed performance for robot manipulators. ISA Transactions. 2019;**91**:41-51

[16] Bechlioulis CP, Rovithakis GA. Prescribed performance adaptative control for multiinput multioutput affline in the control nonlinear systems. IEEE Transactions on Automatic Control. 2010;**55**(5):1220-1226

[17] Kostarigka AK, Rovithakis GA. Prescribed performance output feedback observer-free robust adaptive control of uncertain systems using neural networks. IEEE Transactions on Systems, Man, and Cybernetics. B: Cybernetics. 2011;**41**(6):1483-1494

[18] Craig JJ. Manipulator dynamics. In: Introduction to Robotics: Mechanics and Control. 3rd ed. USA: Pearson Prentice Hall; 2005. pp. 165-200

[19] Zu WH. Virtual decomposition control – general formulation. In: Virtual Decomposition Control: Toward Hyper Degrees of Freedom Robots. Berlin Heidelberg: Springer-Verlag; 2010. pp. 63-109

[20] Karayiannidis Y, Papageorgiou D, Doulgeri Z. A model-free controller for guaranteed prescribed performance tracking of both robot joint positions and velocities. IEEE Robotics and Automation Letters. 2016;**1**(1):267-274

[21] Hsia TC. A new technique for robust control of servo systems. IEEE Transactions on Industrial Electronics. 1989;**36**(1):1-7

[22] Youcef-Toumi K, Ito O. A time-delay controller for systems with unknown dynamics. Journal of Dynamic Systems, Measurement, and Control. 1990;**112**(1): 133-142

[23] Wang YX, Yu DH, Kim YB. 'Robust time delay control for the DC-DC boost converter,' IEEE Transactions on Industrial Electronics. Sep 2014;**61**(9): 4829-4837

[24] Ba DX, Yeom H, Bae JB. A direct robust nonsingular terminal sliding mode controller based on an adaptive time-delay estimator for servomotor rigid robots. Mechatronics. 2019;**59**: 82-94

[25] Lee J, Chang PH, Jin M. An adaptive gain dynamics for time delay control improves accuracy and robustness to significant payload changes for robots. IEEE Transactions on Industrial Electronics. 2020;**67**(4):3076-3085

[26] Jin M, Lee J, Tsagarakis NG. Model free robust adaptive control of humanoid robots with flexible joints. IEEE Transactions on Industrial Electronics. 2017;**64**(2):1706-1715

[27] Lee JY, Jin M, Chang PH. Variable PID gain-tuning method using backstepping control with time delay estimation and nonlinear damping. IEEE Transactions on Industrial Electronics. 2014;**14**(12):6975-6985

[28] Bechlioulis CP, Rovithakis GA. Robust adaptive control of feedback linearizable MIMO nonlinear systems with prescribed performance. IEEE Trans. Automatic Control. 2008;**53**:2090-2099

[29] Ba DX, Truong DQ, Ahn KK. An integrated intelligent nonlinear control method for pneumatic artificial muscle. IEEE/ASME Trans. on Mechatronics. 2016;**21**(4):1835-1845

[30] San P, Ren B, Ge SS, Lee TH, Liu J. Adaptive neural network control of hard disk drives with hysteresis friction nonlinearity. IEEE Transactions on Control Systems Technology. 2011; **19**(2):351-358

[31] Park J, Sandberg IW. Universal approximation using radial-basis-function networks. Neural Computation. 1991;**3**(2):246-257

[32] Khooban H, Vafamand N, Niknam T, Dragicevic T, Blaabjerg F. Model-predictive control based on Takagi-Sugeno fuzzy model for electrical vehicles delayed model. IET Electric Power Applications. 2017;**11**(5):918-934

[33] Wilamowski BM, Cotton NJ, Kaynak O, Dundar G. Computing gradient vector and Jacobian matrix in arbitrary connected neural networks. IEEE Transactions on Industrial Electronics. 2008;**55**(10):3784-3790

[34] Jung S, Kim SS. Hardware implementation of a real-time neural network controller with a DSP and an FPGA for nonlinear systems. IEEE Transactions on Industrial Electronics. 2007;**54**(1):265-271

[35] Chao F, Zhou D, Lin CM, Yang L, Zhou C, Shang C. Type-2 fuzzy hybrid controller network for robotic systems. IEEE Trans. Cybernetics. 2020;**50**(8): 3778-3792

[36] Wang M, Yang A. Dynamic learning from adaptive neural control of robot manipulators with prescribe performance. IEEE Transactions on Systems, Man, and Cybernetics: Systems. 2017;**47**(8):2244-2255

[37] Li Z, Kang Y, Xiao Z, Song W. 'Human–robot coordination control of robotic exoskeletons by skill transfers,' IEEE Transactions on Industrial Electronics. Jun 2017;**64**(6):5171-5181

[38] Jing C, Xu H, Niu X. Adaptive sliding mode disturbance rejection control with prescribed performance for robotic manipulators. ISA Transactions. 2019;**91**: 41-51

[39] Chen M, Ge SS. Adaptive neural output feedback control of uncertain nonlinear systems with unknown hysteresis using disturbance observer. IEEE Transactions on Industrial Electronics. 2015;**62**(12):7706-7716

[40] Chen M, Shao SY, Yang B. Adaptive neural control of uncertain nonlinear systems using disturbance observer. IEEE Trans. Cybernetics. 2015;**62**(12): 7706-7716

[41] Zhang JJ. State observer-based adaptive neural dynamics surface control for a class of uncertain nonlinear systems with input saturation using disturbance observer. Neural Computing and Applications. 2019;**31**:4993-5004

[42] Zhang L, Li Z, Yang C. Adaptive neural network based variable stiffness control of uncertain robotic systems using disturbance observer. IEEE Transactions on Industrial Electronics. 2017;**64**(3):2236-2245

[43] Li Z, Su C, Wang L, Chen Z, Chai T. Nonlinear disturbance observer-based control design for a robotic exoskeleton incorporating fuzzy approximation. IEEE Transactions on Industrial Electronics. 2015;**62**(9):5763-5775

[44] He W, Sun Y, Yan Z, Yang C, Li Z, Kaynak O. Disturbance observer-based neural network control of cooperative multiple manipulators with input saturation. IEEE Trans Neural Networks and Learning Systems. 2020;**31**(5): 1735-1746

[45] Ba DX, Truong DQ, Bae JB, Ahn KK. An effective disturbance-observer-based nonlinear controller for a pump-controlled hydraulic system. IEEE/ ASME Trans. on Mechatronics. 2020; **25**(1):23-32

[46] Ba DX, Bae JB. A precise neural-disturbance learning controller of constrained robotic manipulators. IEEE Access. 2021;**9**:50381-50390

[47] Ba DX. An intelligent sliding mode controller of robotic manipulators with output constraints and high-level adaptation. International Journal of Robust and Nonlinear Control. 2022; **32**(12):6888-6912

Chapter 5

A Dexterous Workspace Optimization for Ten Different Types of General Stewart-Gough Platforms

Burak Inner and Serdar Küçük

Abstract

In this chapter, a dexterous workspace optimization is performed for ten different types of 6-Degrees-Of-Freedom (DOF) General Stewart-Gough Platforms (GSPs). The optimization aims to find the optimum radius of the circumferential circle and separation angles between adjacent vertices of base and moving platforms in order to maximize both the dexterities and workspaces of the manipulators subject to geometric constraints. Particle Swarm Optimization (PSO), increasingly being applied to engineering applications, is used as the optimization algorithm. Finally, the optimization results for ten different types of 6-DOF GSPs are compared to each other in terms of kinematic performances.

Keywords: general Stewart-Gough platforms, dexterous workspace optimization, particle swarm optimization, kinematic performance, stroke length, workspace

1. Introduction

Parallel robot manipulators are closed-loop mechanisms where all links are connected to the base and moving platform at the same time. They have potential advantages over serial robot manipulators such as high rigidity, compact size, high load capacity, fast response, and high precision [1–4]. Basically, parallel manipulators can be classified into two main categories, namely planar and spatial parallel manipulators. The first category composes of planar parallel manipulators which have simple structures and translate along x- and y-axes, and rotate around the z-axis, only. The second category includes spatial parallel manipulators that have 3 to 6-DOF, and can translate and rotate in the three-dimensional space. Stewart-Gough Platforms included in the second group are receiving increased interest from the robotics community and industry recently. They have been used in many potential applications such as multi-axis machine tools earthquake simulators, solar panels, radar antennas, telescopes, walking machines, micromanipulators, and surgery operations [5–7].

Several authors provided valuable contributions to the dimensional optimization of Stewart-Gough Platforms. Some important publications can be given as follows. Pittens and Podhorodeski [8] studied optimizing the local dexterity of a small group of

Stewart platform manipulators. Stoughton and Arai [9] designed a novel structure for the Stewart Platform manipulator. They also optimized the new structure considering both dexterity and workspace volume, and compared the novel structure with the traditional Stewart-Gough Platform in terms of dexterity. Du Plessis and Snyman [10] presented an optimization method for determining the dexterous workspaces of parallel manipulators. They applied the new method for the computation of dexterous workspaces of both planar and spatial Stewart-Gough platforms. Su et. al. [11] conducted a study about optimizing the structural characteristics of the Stewart platform for a large spherical radio telescope using genetic algorithms. They used the condition number of the Jacobian matrix as the objective function and radius of the base platform, adjacent actuator attachment points, and the distance between base and moving platforms as optimization variables. Yao [12] et al. performed the dimensional optimization of the Stewart-Gough platform for a five-hundred-meter aperture spherical radio telescope. The operability and accuracy of the Stewart-Gough platform are the main design objectives of their optimization problem. Mishra and Omkar [13] used different types of evolutionary algorithms (such as particle swarm optimization, genetic algorithm, variants, and simulated annealing) to present a model for singularity analysis of a 6-DOF Stewart-Gough Platform manipulator for precision and surgery. Jiang [14] completed a Ph.D. dissertation about singularity analysis and geometric optimization of two different kinds of parallel mechanisms namely, planar 3-RPR and spatial Stewart-Gough parallel manipulators. An algorithm for optimizing the geometric parameters is developed in order to maximize the singularity-free orientation workspace of the Stewart-Gough platform taking leg length ranges into account. Jiang and Gosselin [15] analyzed the effects of the orientation angles on the singularity-free workspace of the 3x3 Gough-Stewart platform in order to determine the optimal orientation. The same authors also studied the effects of the geometric parameters on the singularity-free workspace in order to determine the optimal architecture for the minimal simplified symmetric manipulator 3x3 Gough-Stewart platform [16]. Furthermore, they analyzed the maximal singularity-free total orientation workspace of the 3x3 Gough-Stewart platform [17].

Although there have been several studies about the optimization of Stewart-Gough platform manipulators in the literature, most of these studies have been restricted to 6-legged 3x3, 6x3, and 6x6 Stewart platform manipulators only. The study in this chapter presents a dexterous workspace optimization of all possible types of 6-legged architectures having five different prismatic active actuator stroke lengths between the base and moving platforms. These architectures consist of commonly used ten different types of Stewart-Gough platforms, namely the 6-DOF 3x3, 4x3, 4x4, 5x3, 5x4, 5x5, 6x3, 6x4, 6x5, and 6x6 parallel mechanisms as shown in **Figure 1**. The dexterities and workspaces of the 6-DOF Stewart-Gough platforms are used as the objective function subject to geometric constraints. The radius of the circumferential circles and separation angle between adjacent vertices of the base and moving platforms are considered as the optimization variables. A Particle Swarm Optimization (PSO) algorithm which is increasingly being applied in various engineering applications (wireless sensor networks [18], electromagnetics [19], biomedical [20], electronics [21], control [22], and robotics [23]), is used as the optimization tool. In general, a PSO algorithm can be implemented easily and it is computationally inexpensive in terms of both memory requirements and CPU time. It has been proven to be effective for especially dynamic optimization problems with multi-dimensional search spaces [24–28]. Finally, the optimization results for ten different types of 6-DOF GSPs with five different actuator stroke lengths are compared to each other and ranked in terms of kinematic performances.

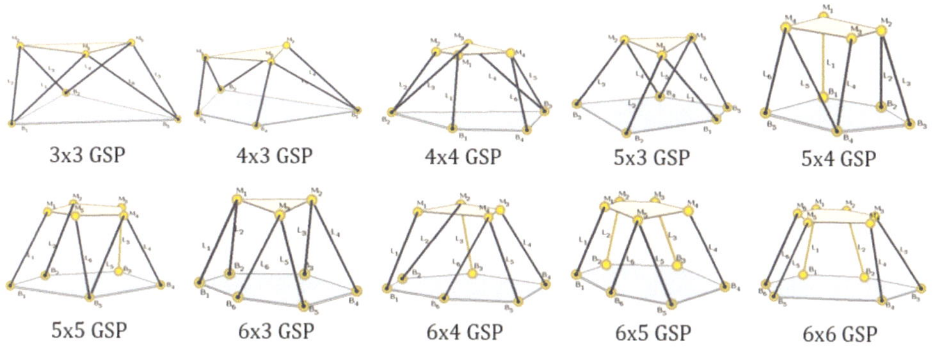

Figure 1.
6-DOF 3x3, 4x3, 4x4, 5x3, 5x4, 5x5, 6x3, 6x4, 6x5, 6x6 Stewart-Gough platforms.

2. Background

Geometric description, inverse kinematics, and Jacobian matrix derivations of 6-legged GSPs are explained in this section.

2.1 Geometric description of 6-legged GSPs

The possible structures for Stewart-Gough platform mechanisms with 6-legs are constructed by providing six active actuators between the base and moving platforms and 12 passive joints attached to both ends of each active actuator [14]. While the passive joints might be selected as spherical or universal joints, the active actuators are chosen as prismatic joints only. As stated in the introduction Section, the ten different types of 6-legged Stewart-Gough platform mechanisms can be constructed by changing the placements of the attachment points on the fixed base and moving platforms. For instance, in order to construct a 6-legged 3x3 Stewart platform shown in **Figure 2a**, the six legs are connected to both moving and base platforms at three attachment points while a 6-legged 6x6 Stewart-Gough platform illustrated in **Figure 2b** requires six attachment points at the base and moving platforms. In general to perform kinematic and dynamic operations, two reference coordinate frames namely B={X, Y, Z} and M={x, y, z} are attached to the centers of the base (O) and moving platforms (P), respectively. The passive joints are connected to B_i and M_i attachment points (i=1, 2, 3 for 3x3 of GSP and i=1, 2 ... 6 6x6 of GSP) on the fixed base and moving platform, respectively. The $B_i = \begin{bmatrix} b_{ix} & b_{iy} & b_{iz} \end{bmatrix}^T$ and $M_i = \begin{bmatrix} m_{ix} & m_{iy} & m_{iz} \end{bmatrix}^T$ are the position vectors of the points B_i and M_i in the B and M coordinate systems, respectively. The ψ_{b_i} and ψ_{m_i} illustrate the separation angles between adjacent vertices (i=1, 2, 3 for 3x3 of GSP and i=1, 2 ... 6 6x6 of GSP) of base and moving platforms, respectively.

2.2 Inverse kinematics and Jacobian matrix

The inverse kinematics of the GSPs can be determined by using the following equation when the position$P = \begin{bmatrix} p_x & p_y & p_z \end{bmatrix}^T$ and orientation matrix$R_{XYZ}(\alpha, \beta, \gamma)$ of the end-effector in terms of base coordinate frames are given as

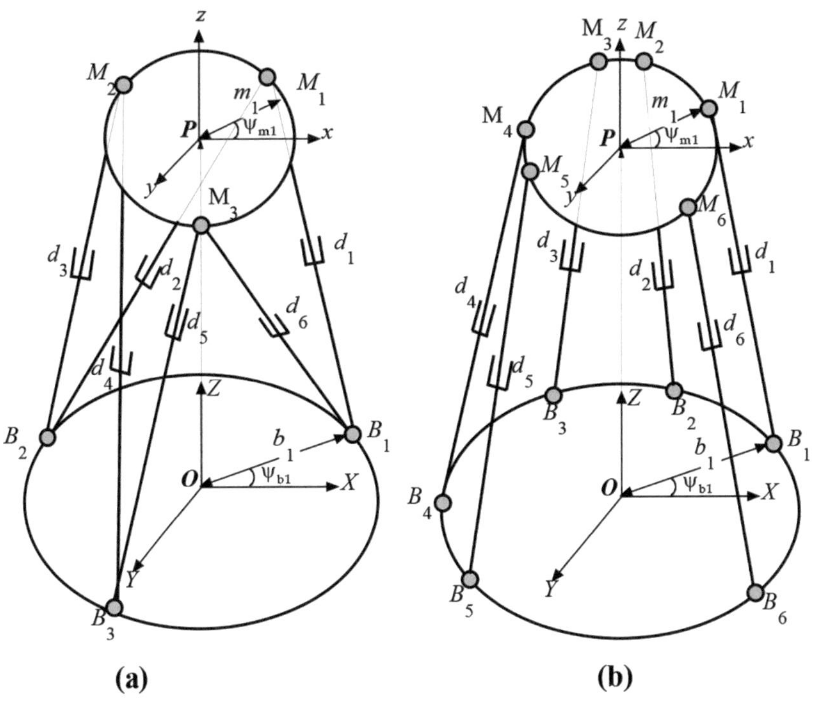

Figure 2.
(a) The 6-legged 3x3 of GSP. (b) The 6-legged 6x6 of GSP.

$$d_i = R_{XYZ}M_i + P - B_i, i = 1,2,\cdots,6 \tag{1}$$

where d_i denotes the active link lengths. The Jacobian matrix can be derived by applying the loop-closure equation to each limb.

$$\overrightarrow{OP} + \overrightarrow{PM_i} = \overrightarrow{OB} + \overrightarrow{B_iM_i}, i = 1,2,\cdots,6 \tag{2}$$

where i equals the number of limbs. Differentiating Eq. (2) with respect to the time and eliminating the angular velocity of the active actuators with respect to the base frame in the resultant equation, the following identity is found [1].

$$z_i \bullet v_p + (m_i \times z_i) \bullet \omega_p = \dot{d}_i, i = 1,2,\cdots,6 \tag{3}$$

where v_p, ω_p and $\dot{d}_i$ are the linear and angular velocities of the moving platform and the linear velocities of active actuators, respectively. The term z_i denotes the unit vector alongB_iM_i. Eq. (3) can be stated as

$$J_A \bullet \dot{\boldsymbol{x}} = J_B \bullet \dot{\boldsymbol{q}}, i = 1,2,\cdots,6 \tag{4}$$

where $\dot{\boldsymbol{x}} = \begin{bmatrix} v_p & \omega_p \end{bmatrix}^T = \begin{bmatrix} v_{p_x} & v_{p_y} & v_{p_z} & \dot{\alpha} & \dot{\beta} & \dot{\gamma} \end{bmatrix}^T$, $\dot{\boldsymbol{q}} = \begin{bmatrix} \dot{d}_1, \dot{d}_2, \cdots, \dot{d}_6 \end{bmatrix}^T$, J_B is equal to the 6x6 identity matrix and J_A is as follows.

$$J_A = \begin{bmatrix} z_1^T & m_1 \times z_1 \\ z_2^T & m_2 \times z_2 \\ z_3^T & m_3 \times z_3 \\ z_4^T & m_4 \times z_4 \\ z_5^T & m_5 \times z_5 \\ z_6^T & m_6 \times z_6 \end{bmatrix} \tag{5}$$

The overall Jacobian matrix is obtained as

$$J = J_B^{-1} J_A \tag{6}$$

3. Optimization constraints

The optimization problem is evaluated by taking some important geometric constraints into account. The geometric constraints include the minimum and maximum radii of the circumferential circles and separation angles between adjacent vertices which determine the connection points of the legs both on the base and the moving platforms, respectively.

The first geometric constraint is the minimum and maximum radii of the circumferential circles of the base and moving platforms. The minimum limits of the circumferential circles can be obtained by considering the physical dimensions of the passive joints (such as universal and spherical joints). **Figure 3** illustrates the placements of the passive joints on the base and moving platform for providing the physically minimum radii of the circumferential circles. The radius of the joints and the distance between two consecutive passive joints are denoted as r_i and j_m, respectively. The minimum limits of the circumferential circles ($r_{b(min)}$)) depicted in **Figure 3** are determined by considering the radius of passive universal joints (r_j) and the distance between two consecutive passive joints (j_m). **Table 1** gives the minimum limits of the circumferential circles of the base and moving platform for 3 to 6-legged GSP mechanisms. The maximum radius of the circumferential circles of the base ($r_{b,max}$) and moving ($r_{m,max}$) platforms can be chosen based on the design requirements.

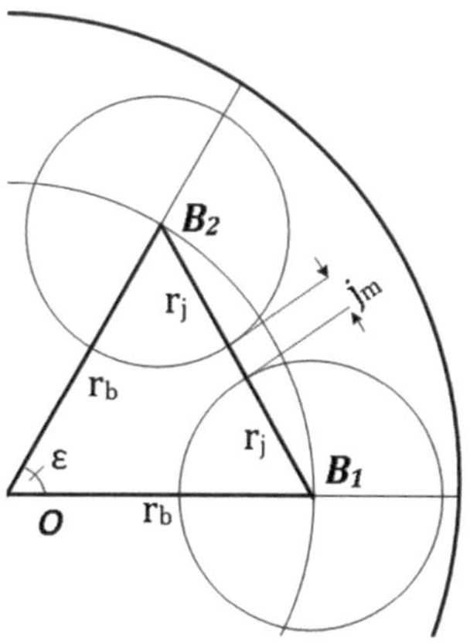

Figure 3.
The minimum radius of the circumferential circles.

Number of connection points	3	4	5	6
Maximum angle (ε)	120°	90°	72°	60°
Minimum radius of the platform	$\frac{2r_j+j_m}{2\sin(60)}$	$\frac{2r_j+j_m}{2\sin(45)}$	$\frac{2r_j+j_m}{2\sin(36)}$	$\frac{2r_j+j_m}{2\sin(30)}$

Table 1.
Maximum separation angle for a minimum radius of the base and moving platforms.

The second geometric constraint is the separation angles between adjacent vertices which determine the connection points of the legs both on the base and moving platforms, respectively. The connection points refer to the centers of the passive joints whose coordinates are denoted as B_i on the base and M_i on the moving platforms, respectively. The minimum separation angle depends on the radius of the circumferential circles of the base or moving platforms. Since the radius of the passive joints is constant, as the radius of the circumferential circle gets larger, the separation angle becomes smaller as shown in **Figure 4**. The minimum separation angle for the base and moving platform is computed by using the cosines theorem on OB_1B_2 and PM_1M_2 triangle where the radius of the base and moving platforms denotes as r_b and r_m, respectively. $\delta_{b,min}$ and $\delta_{m,min}$ can be extracted as

$$\delta_{bi,min} = cos^{-1}\left(1 - \frac{(2r_j + j_m)^2}{2r_b^2}\right) \quad i = 1,2,\cdots,6 \tag{7}$$

$$\delta_{mi,min} = cos^{-1}\left(1 - \frac{(2r_j + j_m)^2}{2r_m^2}\right) \quad i = 1,2,\cdots,6 \tag{8}$$

The following statement can be easily written since Eqs. (7) and (8) are the common identities for the minimum separation angles on the base and moving platforms, respectively.

$$\delta_{bi+1,min} = \delta_{bi,min} \; i = 1,2,\cdots,5. \tag{9}$$

$$\delta_{mi+1,min} = \delta_{mi,min} \; i = 1,2,\cdots,6 \tag{10}$$

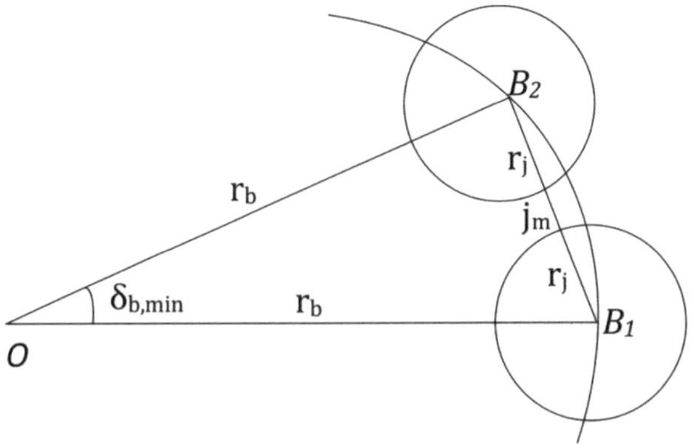

Figure 4.
The illustration of the minimum separation angles.

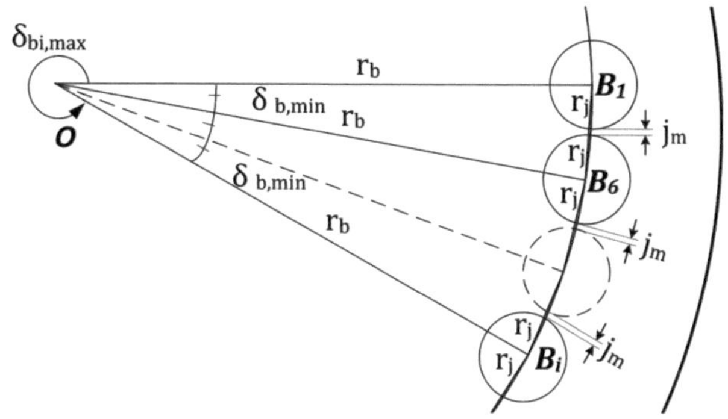

Figure 5.
The illustration of the maximum separation angles.

Number of legs	3	4	5	6
Maximum angle (ε)	$360-(4-i)\delta_{b,min}$	$360-(5-i)\delta_{b,min}$	$360-(6-i)\delta_{b,min}$	$360-(7-i)\delta_{b,min}$

Table 2.
The maximum separation angles of the base and moving platforms for GSPs.

The maximum separation angles of the base ($\delta_{bi,max}$) and moving ($\delta_{mi,max}$) platforms can be found considering the **Figure 5**. **Table 2** illustrates the maximum separation angles of the base ($\delta_{bi,max}$) and moving ($\delta_{mi,max}$) platforms for 3 to 6 legged GSPs. The separation angles ($\delta_{bj},\delta_{mj,}$) of the base and moving platforms are randomly chosen between the minimum angle denoted by Eqs. (7) and (8), and the maximum angle stated by **Table 2**, respectively where j is the number of legs that vary between 3 and 6.

4. Optimization objectives

In this optimization problem, the dexterities and workspaces of the manipulators are used as the optimization objectives subject to geometric constraints. Dexterity is a very important measure that directly affects the kinematic performance of robotic manipulators. The capability of achieving small displacements in arbitrary directions in the manipulator workspace [29] can be described as dexterity. There are some local and global dexterity measures based on the Jacobian matrix [30–32]. It should be noted that dexterity based on the Jacobian matrix may not directly be computed due to the dimensional inconsistencies of the matrix elements. Therefore a characteristic length is determined to homogenize the elements of the Jacobian matrix since the first three columns of the matrix J_A given by Eq. (5) have the units of length, whereas the last three columns have the units of length^2.

$$J_{AH} = \begin{bmatrix} J_{A1} & \frac{1}{L} J_{A2} \end{bmatrix} \tag{11}$$

where J_{AH} is the homogenized Jacobian matrix of J_A. The J_{A1} and J_{A2} are the 6x3 submatrices. The characteristic length L is determined as the radius of the moving platform. Thus the homogenized overall Jacobian matrix J_H is described as

$$J_H = J_B^{-1} J_{AH} \tag{12}$$

The local dexterity based on the condition number (κ) of the Jacobian matrix is given by

$$\kappa = \|J_H\| \|J_H^{-1}\| \tag{13}$$

where $\|\|$ illustrates the matrix norm as

$$\|J_H\| = \sqrt{tr(J_H n J_H^T)} \tag{14}$$

where n is a diagonal matrix. Condition number κ changes between 1 and ∞. Inverse condition number $\eta = 1/\kappa$ limited between 0 and 1 is used in general to measure dexterity easily. The inverse condition number illustrates the local behavior of the manipulator. In order to measure the global property of the manipulator, **G**lobal **D**exterity **I**ndex (GDI) is used as

$$GDI = \frac{\int_W \eta dW}{\int_W dW} \tag{15}$$

where the denominator of Eq. (15) illustrates the workspace volume of the manipulator. As the GDI approaches unity the manipulator gains better gross motion capability. The following identity can be used for GDI due to avoiding the troubles while computing the integrals in Eq. (15).

$$GDI = \frac{\sum \eta}{nmp} \tag{16}$$

where the nmp illustrates the number of points in the workspace and the numerator shows the sum of η values in the workspace grids.

Finally, the optimization problem for the ten different types of GSPs with five different actuator stroke lengths can be stated as the maximization of both the dexterities and workspaces of the manipulators subject to geometric constraints.

$$\text{Max GDI and Worksapce} \tag{17}$$

$$\text{Subject to}$$

$$r_{b,min} \le r_b \le r_{b,max}$$

$$r_{m,min} \le r_m \le r_{m,max}$$

$$\delta_{bi,min} \le \delta_{bi} \le \delta_{bi,max} \quad i = 1,2,\cdots,6.$$

$$\delta_{mi,min} \le \delta_{mi} \le \delta_{mi,max} \quad i = 1,2,\cdots,6.$$

5. The particle swarm optimization

PSO is a robust stochastic optimization algorithm inspired by the biological social behavior of a swarm of birds or a school of fish. PSO was firstly introduced by

Kennedy and Eberhart in 1995 [33] while they have been attempting to simulate the motion of bird swarms. PSO is capable of finding optimal or near-optimal solutions in shorter computation time and also suitable for searching in large search space. Potential solutions within the search space are called particles and a population-based search is performed by considering the fitness values that are obtained from the positions of the particles. At each flight cycle, particles fly around in a multidimensional search space with a velocity, and the objective function is evaluated for each particle based on its position. Velocity directing the flight of particles is updated based on the particle's current velocity, the particle's own best fitness value and the global best fitness value of any particle in the population. Thus the movement of each particle is guided toward the local and the best-known positions in the search space. This is expected to move the swarm toward the best solutions.

PSO is started with a random population referred to as a swarm and search optima by updating generations iteratively. Each particle in the swarm is treated as a point in an N-dimensional search space and keeps track of the best solution (fitness) which has been achieved by that particle so far. The best solution is called as personal best (Pbest). The best value achieved up to now by any particle in the population is called as global best (Gbest). The basic concept of PSO lies in accelerating each particle toward its Pbest and the Gbest locations. The velocity of each particle in the swarm is updated by using the following equation. The performance and accuracy of the PSO algorithm are mostly based on the appropriate selection of constriction factor (χ), inertia weight (ω_k), and learning factors (c_1, c_2) parameters where k denotes the iteration number.

6. Simulation results

In this study, the kinematic structures of the ten GSPs are optimized for five different linear actuator stroke lengths. **Figure 6** illustrates a linear actuator whose extended length composes of stroke and retracted lengths. **Table 3** illustrates the joint radius (r_j), the stroke, retracted, and extended lengths of the linear actuators commonly used in industrial applications. According to **Table 3** the actuator stroke lengths change between 50mm and 250mm. In this study, the actuator stroke lengths are selected as 50mm, 100mm, 150mm, 200mm, and 250 mm, which are the common lengths used in industry. The radii of the passive joints that change between 13.5mm and 30mm with respect to the **Table 3** are selected as 18mm for five different stroke lengths. Finally, the distance between two consecutive passive joints is selected as 6mm in accordance with the commercial PI M^{-840}.PD3 6-axis hexapod.

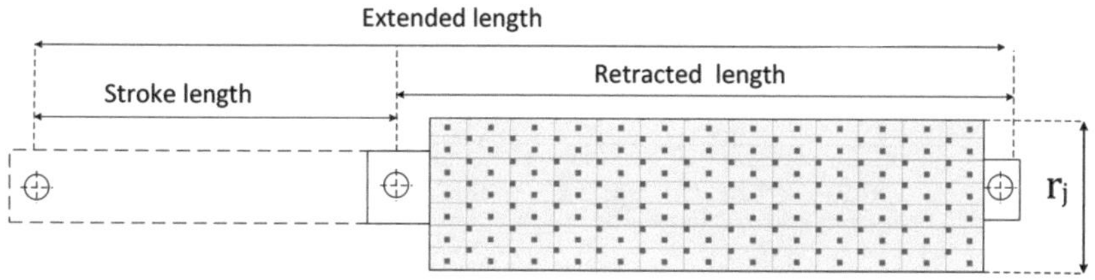

Figure 6.
Schematic diagram of a linear actuator.

Linear actuator	Model	Stroke length (mm)	Retracted length (mm)	Extended length (mm)	r_i (mm)
Oriental motor	DRL60PA4-05G	50	133	183	30
	DRL60PB4-10G	100	186.5	286.5	30
PI	M^{-235}.5DG	50	218	268	13.5
Linear-Mech	LMI 02-C100	100	345	445	22.5
	LMI 02-C150	150	395	545	22.5
	LMI 02-C200	200	445	645	22.5
	LMI 02-C250	250	495	745	22.5
	LMP 03-C100	100	252	352	25
	LMP 03-C150	150	302	452	25
	LMP 03-C200	200	352	552	25
	LMP 03-C250	250	402	652	25

Table 3.
The stroke and retracted lengths of the linear actuators commonly used in industrial applications.

The three-dimensional Cartesian workspace volumes are planned such that the end-effectors of the ten GSPs with five different linear actuator stroke lengths can easily reach their own workspaces with the reference orientation angle $R_{XYZ}(\alpha, \beta, \gamma)$. **Table 4** illustrates the travel ranges of the orientation angles for the GSPs that have been generally used for industrial applications. The orientation angles of the moving platform are selected as $\alpha = \beta = \gamma = 0$ since the travel ranges of the commercial GSPs summarized in **Table 4** vary between very small intervals.

The minimum and maximum separation angles on the base and moving platform are the variables that can be calculated by using Eqs. (7), (8), and **Table 2**. In addition, **Table 5** illustrates the values of the minimum and maximum radii for base and moving platforms. The value of minimum radii for the base and moving platforms are computed as 42 by using **Table 1** while the maximum radii of the base and moving platforms given in **Table 5** are used with respect to the stroke lengths of the GSPs.

In this optimization problem, the individuals of a swarm are evaluated by considering the maximization of the objective functions stated by Eq. (17). For each generation, dexterities and workspaces of the ten different GSPs with five different linear

Manipulator	Travel range			Manipulator	Travel range		
	θ_x	θ_y	θ_z		θ_x	θ_y	θ_z
Newport HXP50	±9°	±8.5°	±18°	PI M-850KHLH	±3°	±3°	±4°
PI H-824	±7.5	±7.5	±12.5	PI M-850KHLAH	±5°	±5°	±5°
PI M-811	±10°	±10°	±21°	SYMETRIE Bora	±15°	±15°	±15°
PI M-824.3DG	±7.5°	±7.5°	±12.5°	SYMETRIE Breva	±15°	±15°	±15°
PI M-824.3PD	±7.5°	±7.5°	±12.5°	SYMETRIE Sonora	±2°	±2°	±2°

Table 4.
Orientation angles of GSPs used for industrial applications.

	Stroke 50 (in mm)		Stroke 100 (in mm)		Stroke 150 (in mm)		Stroke 200 (in mm)		Stroke 250 (in mm)	
	r_b	r_m	r_b	r_m	r_b	r_m	r_b	r_m	r_b	r_m
min	42	42	42	42	42	42	42	42	42	42
max	210	210	420	420	630	630	840	840	1050	1050

Table 5.
The values of the minimum and maximum radii for base & moving platforms.

actuator stroke lengths are computed based on the radius of the circumferential circle, separation angles, and constant orientation angles of $\alpha = \beta = \gamma = 0$. The parameters of the PSO algorithm are selected as$\chi = 0.7298$, $c_1 = c_2 = 2.05$. The population size is selected as 40 particles. Each particle has composes of 13 elements. The objective function is evaluated for 60 generations.

The optimization results of ten different GSPs with five different linear actuator stroke lengths are shown in tables. Due to the page limitations, the optimization results of five GSPs are given only (**Tables 6–12**). The tables include the radius of the circumferential circles, separation angles between adjacent vertices, workspace volumes (WSP), GDI values, shapes of the base and moving platforms. The base and moving platforms are plotted in the same axes for easy comparison and illustrated as red and blue colors in the figures, respectively. The connection points of the legs on base and moving platforms, retracted and extracted lengths are also given in tables in order to show the data belonging to the manipulators in a compact form.

As can be seen in **Tables 6–12**, the radii of the base platforms are optimized larger than the radii of the moving platforms for each actuator stroke length of the ten different GSPs. The optimization results show that the locations of the connection points on the base platform form roughly a triangle. As the GSP mechanism with four connection points is taken into account, the location of the two connection points is optimized separately, while the locations of the last two consecutive connection points are optimized close to each other. These two close consecutive connection points can be considered as one connection point. Thus, a rough triangle forms with the first two connection points and the last two close consecutive connection points. There are one separate connection point and two close consecutive connection points for GSP mechanisms with five connection points that form a rough triangle base platform. Finally, the GSP mechanisms with six connection points form a rough triangle having three close consecutive connection points.

Most of the connection points on the moving platforms of the GSPs are located separately while the connection points on the moving platforms of the 4x4, 5x5, 6x5, and 6x6 GSP mechanisms are optimized as a roughly triangle like in the base platforms. The base and moving platforms of 6x5 and 6x6 GSPs are optimized as a rough equilateral triangle for each actuator stroke length. As can be seen in **Tables 6–12**, the GDI values of 6x5 and 6x6 GSPs are higher than the others. It can be concluded that the rough equilateral triangle structure for the base and moving platforms can produce better kinematic performance.

It can be concluded from **Tables 6–12** that as the actuator stroke lengths get longer the workspace of the GSPs become larger. However, the GDI values of the GSPs do not continuously increase as the actuator stroke lengths get longer. **Figure 7** illustrates the GDI values of the ten different GSPs for each actuator stroke length. As can be

Optimized design variables		Stroke lengths (in mm)				
		50	100	150	200	250
Radius (in mm)	r_b	130.79	203.35	42.05	404.34	515.41
	r_m	45.12	42.31	52.91	54.56	58.66
Separation angles of the moving platform (in degrees)	δ_{m1}	0.00	0.00	172.60	92.54	93.93
	δ_{m2}	121.13	118.83	293.20	211.15	212.92
	δ_{m3}	239.13	238.69	105.57	314.73	318.05
Separation angles of the base platform (in degrees)	δ_{b1}	36.16	38.85	105.57	0.00	0.00
	δ_{b2}	159.30	158.09	224.28	113.46	113.97
	δ_{b3}	275.31	279.59	345.77	230.19	231.21
Retracted Length (in mm)		150	250	280	592	740
Extended Length (in mm)		200	350	430	792	990
GDI		0.918	0.829	0.821	0.784	0.785
WSP (in cm^3)		116.28	328.23	1133	1344.4	1550

Connection points of the legs on base and moving platforms						
Legs	L_1	L_2	L_3	L_4	L_5	L_6
Base	B_1	B_1	B_2	B_2	B_3	B_3
Moving	M_3	M_1	M_1	M_2	M_3	M_1

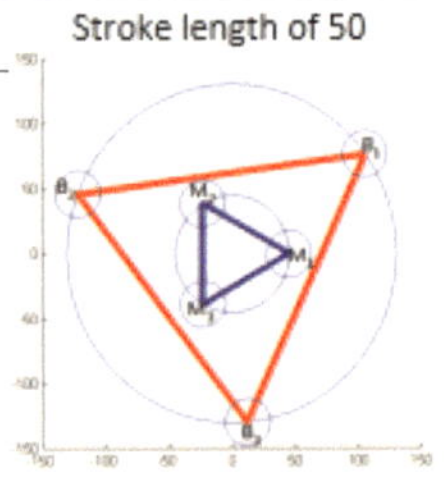

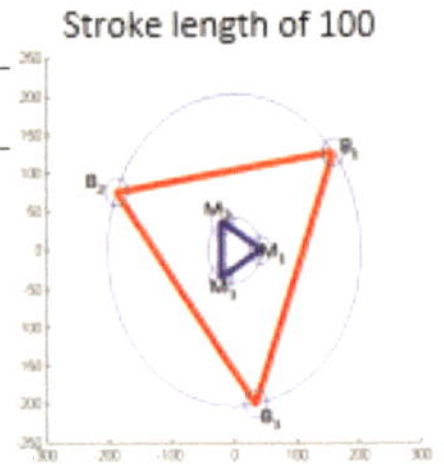

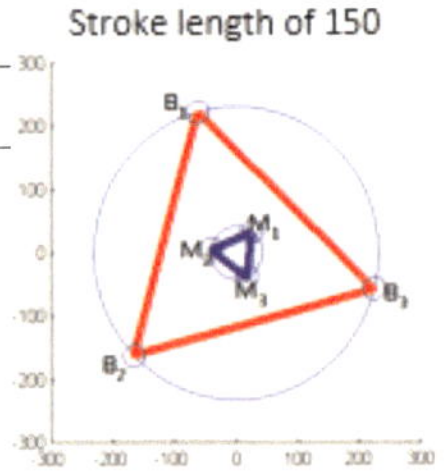

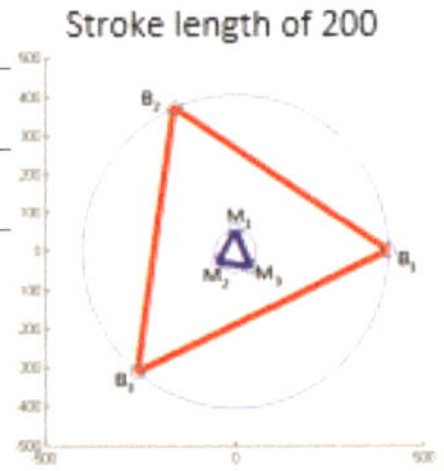

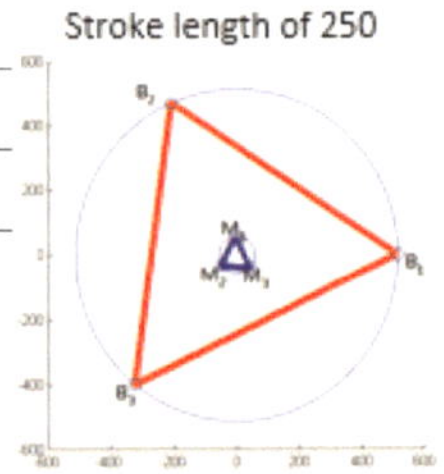

Table 6.
The optimization results for 3x3 GSP.

seen in **Figure 7** the actuator stroke length of 100mm has better GDI values for ten different GSPs in general. It can be noticed that the actuator stroke length of 150mm has very close GDI values to the actuator stroke length of 100mm. Moreover, the actuator stroke length of 150mm has a larger workspace than the actuator stroke

Optimized design variables		Stroke lengths (in mm)					
		50	100	150	200	250	
Radius	r_h	154.75	200.28	231.15	411.20	520.90	Stroke length of 50 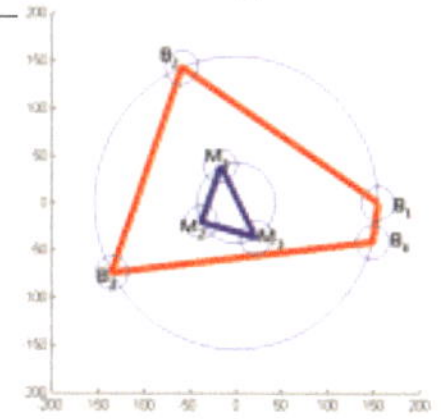
	r_m	42.81	42.11	42.02	42.54	42.15	
Separation angles of the moving platform (in degrees)	δ_{m1}	113.76	72.44	67.90	67.28	64.34	Stroke length of 100 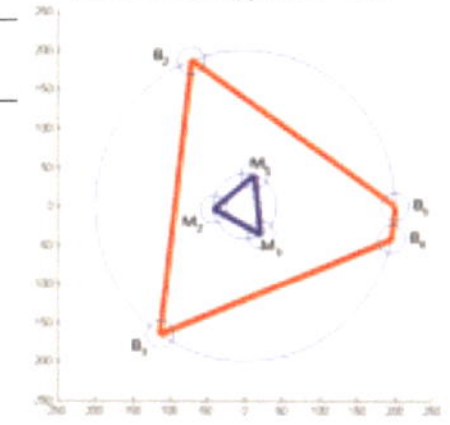
	δ_{m2}	207.78	185.25	185.13	187.36	184.83	
	δ_{m3}	301.16	300.18	300.03	300.84	300.22	
Separation angles of the base platform (in degrees)	δ_{b1}	0.00	0	0	0	0	Stroke length of 150 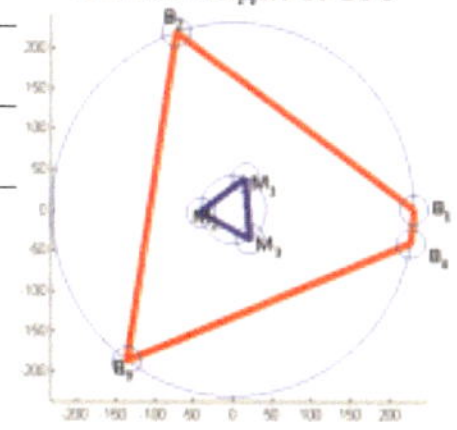
	δ_{b2}	112.55	110.98	108.54	114.77	114.45	
	δ_{b3}	208.72	235.65	233.95	236.84	236.28	
	δ_{b4}	344.40	347.96	349.57	354.14	355.38	
Retracted Length (in mm)		150	250	280	592	740	Stroke length of 200 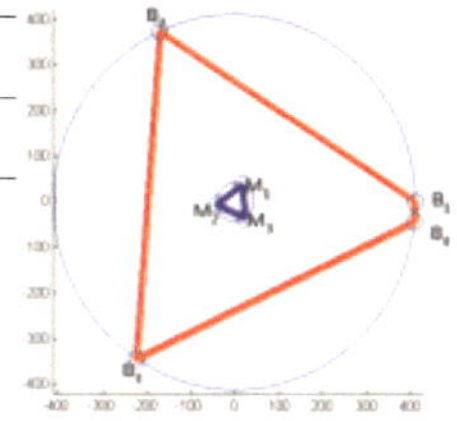
Extended Length (in mm)		200	350	430	792	990	
GDI		0.745	0.829	0.821	0.784	0.785	
WSP (in cm^3)		55.66	328.23	1133	1344.36	1550	

Connection points of the legs on base and moving platforms							
Legs	L_1	L_2	L_3	L_4	L^5	L_6	Stroke length of 250 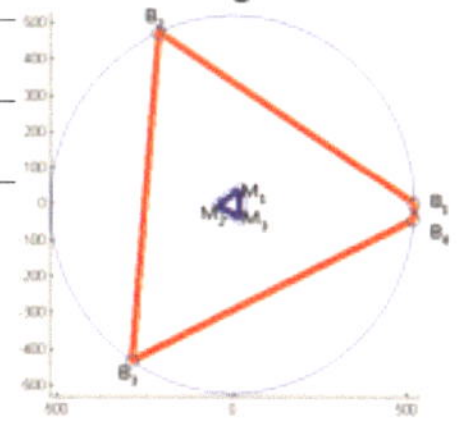
Base	B_1	B_2	B_2	B_3	B_3	B_4	
Moving	M_1	M_1	M_2	M_2	M_3	M_3	

Table 7.
The optimization results for 4x3 GSP.

length of 100mm. Thus, the designers can select the actuator stroke lengths of 100m or 150mm considering the workspace requirements to construct GSPs with higher kinematic performances.

Optimized design variables		Stroke lengths					
		50	100	150	200	250	
Radius	r_b	157.13	204.43	234.16	406.02	528.65	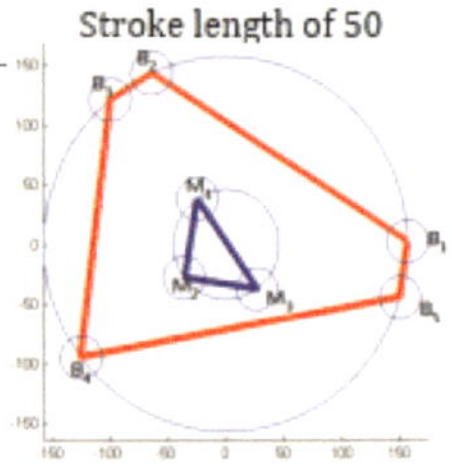
	r_m	45.85	42.13	42.08	42.00	42.08	
Separation angles of the moving platform (in degrees)	δ_{m1}	123.13	70.68	70.18	64.01	64.48	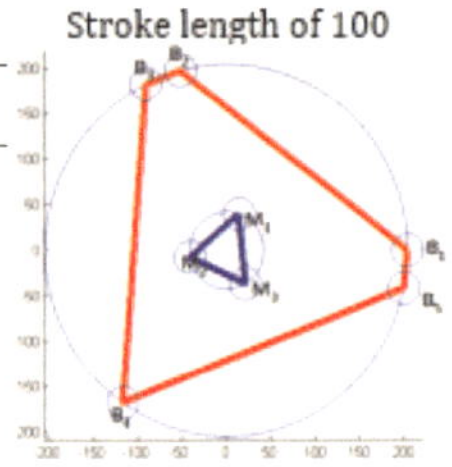
	δ_{m2}	216.98	190.25	189.85	185.09	186.00	
	δ_{m3}	305.48	300.19	300.12	299.97	300.12	
Separation angles of the base platform (in degrees)	δ_{b1}	0.91	0.00	0.00	0.00	0.00	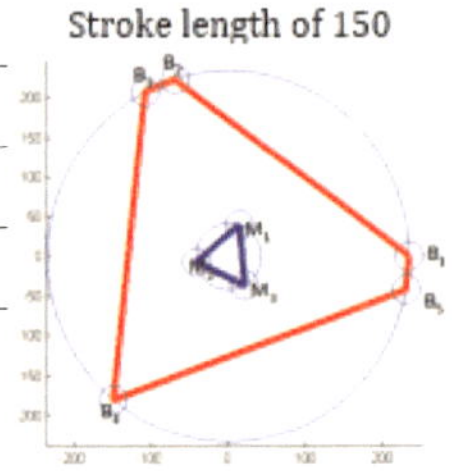
	δ_{b2}	114.22	104.85	107.22	114.38	117.46	
	δ_{b3}	129.58	116.64	117.51	120.31	122.01	
	δ_{b4}	217.21	234.90	230.17	237.56	240.87	
	δ_{b5}	343.54	348.20	349.71	354.07	355.36	
Retracted Length (in mm)		150	250	280	592	740	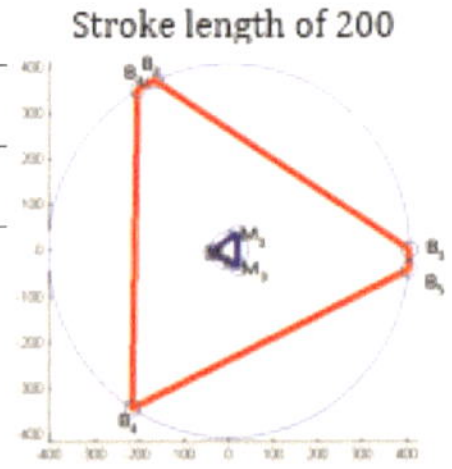
Extended Length (in mm)		200	350	430	792	990	
GDI		0.705	0.820	0.811	0.772	0.785	
WSP (cm^3)		58.20	290.20	1094.99	1399.52	1466.90	

Connection points of the legs on base and moving platforms							
Legs	L_1	L_2	L_3	L_4	L_5	L_6	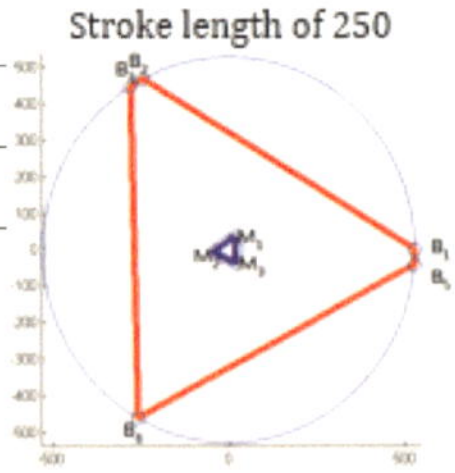
Base	B_1	B_2	B_3	B_4	B_4	B_5	
Moving	M_1	M_1	M_2	M_2	M_3	M_3	

Table 8.
The optimization results for 5x3 GSP.

The kinematic performances of the ten different GSPs are compared by their actuator stroke length. The GDI values of GSPs for each actuator stroke length given by **Table 13** are sorted out from the highest to the lowest. Comparisons show that the best manipulators for each actuator stroke length are 6x5 and 6x6 GSP mechanisms

Optimized design variables		**Stroke lengths**					
		50	**100**	**150**	**200**	**250**	
Radius	r_b	129.45	210.49	234.74	414.76	524.13	Stroke length of 50
	r_m	42.18	44.12	42.31	42.17	42.02	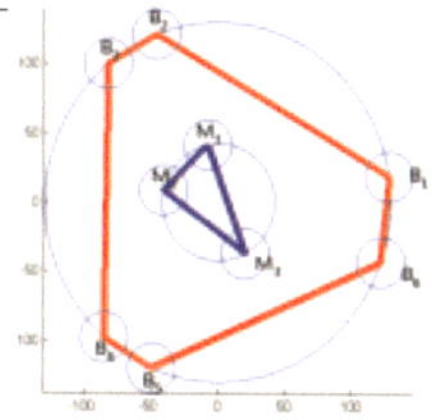
Separation angles of the moving platform (in degrees)	δ_{m1}	100.90	80.31	72.75	68.69	64.91	Stroke length of 100
	δ_{m2}	167.76	194.47	189.30	187.28	185.40	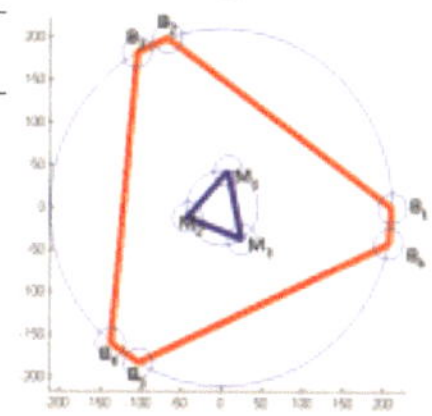
	δ_{m3}	298.10	303.15	300.47	300.26	300.02	
Separation angles of the base platform (in degrees)	δ_{b1}	8.39	0.00	3.34	0.00	0.01	Stroke length of 150
	δ_{b2}	110.98	108.50	115.52	115.16	116.84	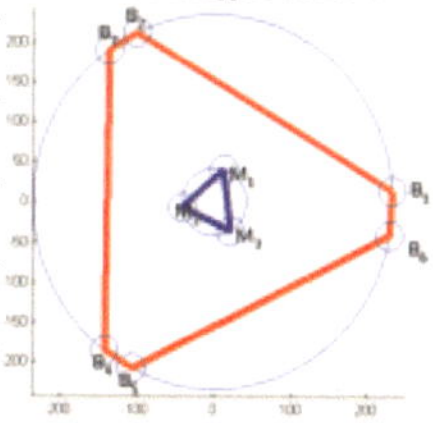
	δ_{b3}	129.65	119.95	125.79	120.97	121.43	
	δ_{b4}	228.59	229.40	232.40	238.17	238.17	
	δ_{b5}	247.26	240.85	242.66	243.98	242.81	
	δ_{b6}	340.51	348.53	349.67	354.04	355.40	
Retracted Length (in mm)		150	250	280	592	740	Stroke length of 200
Extended Length (in mm)		200	350	430	792	990	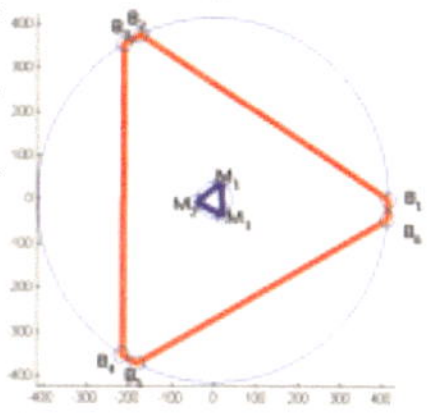
GDI		0.642	0.811	0.791	0.771	0.776	
WSP (cm^3)		56.88	250.09	1099.23	1309.14	1514.48	

Connection points of the legs on base and moving platforms							Stroke length of 250
Legs	L_1	L_2	L_3	L_4	L_5	L_6	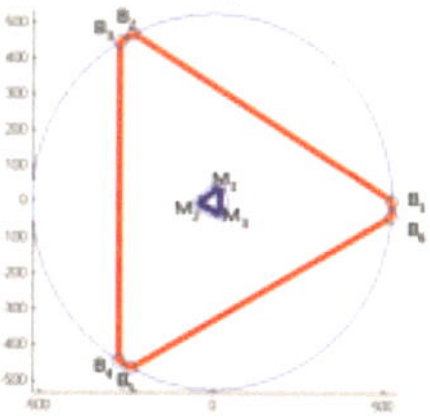
Base	B_1	B_2	B_3	B_4	B_5	B_6	
Moving	M_1	M_1	M_2	M_2	M_3	M_3	

Table 9.
The optimization results for 6x3 GSP.

whose lower and upper bounds of GDI values change between 0.883 and 0.928, respectively. The physical meaning of these values is that the 6x5 and 6x6 manipulator structures provide better dexterous maneuverability and kinematic performance than the others. These manipulators have also larger workspaces than those of the same

Optimized design variables		Stroke lengths				
		50	100	150	200	250
Radius	r_b	168.19	263.47	304.97	494.46	741.16
	r_m	144.04	239.37	268.54	399.81	687.17
Separation angles of the moving platform (in degrees)	δ_{m1}	63.16	65.25	50.65	56.58	47.02
	δ_{m2}	157.35	167.07	160.31	166.37	155.03
	δ_{m3}	260.59	282.24	276.55	283.37	272.38
	δ_{m4}	343.04	349.76	350.91	353.90	344.17
Separation angles of the base platform (in degrees)	δ_{b1}	0.00	16.74	1.91	1.45	2.86
	δ_{b2}	109.09	121.81	114.27	113.35	111.73
	δ_{b3}	198.28	208.17	201.04	215.69	195.59
	δ_{b4}	212.63	217.31	208.93	220.56	198.88
	δ_{b5}	227.02	237.11	228.21	227.64	227.19
	δ_{b6}	322.44	330.06	326.60	339.69	317.50
Retracted Length (in mm)		150	250	280	592	740
Extended Length (in mm)		200	350	430	792	990
GDI		0.864	0.822	0.822	0.781	0.786
WSP (cm^3)		62.56	651	1415	1502	1586

Connection points of the legs on base and moving platforms						
Legs	L_1	L_2	L_3	L_4	L_5	L_6
Base	B_1	B_2	B_3	B_4	B_5	B_6
Moving	M_1	M_2	M_2	M_3	M_4	M_4

Stroke length of 50

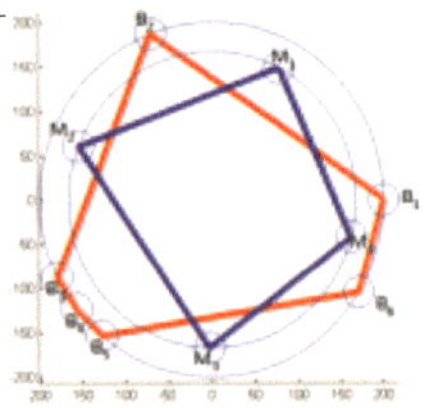

Stroke length of 100

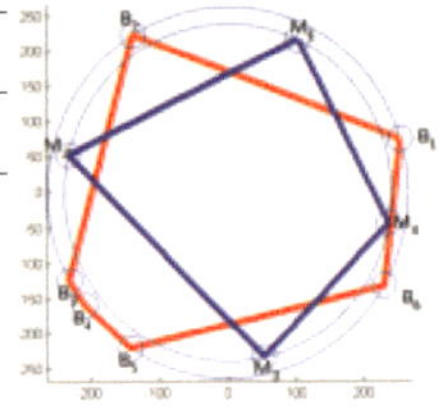

Stroke length of 150

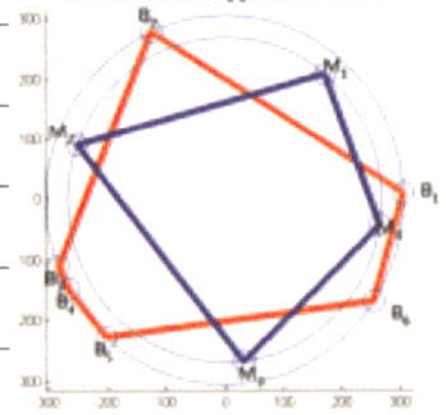

Stroke length of 200

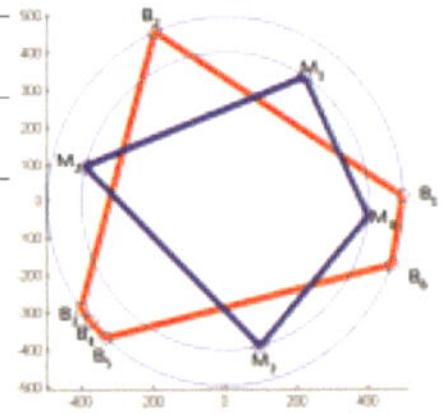

Stroke length of 250

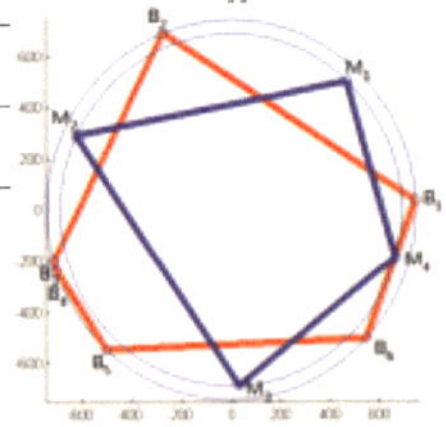

Table 10.
The optimization results for 6x4 GSP.

Optimized design variables		Stroke lengths					
		50	100	150	200	250	
Radius	r_b	209.95	256.18	334.88	561.02	635.39	Stroke length of 50
	r_m	154.31	173.90	168.87	379.88	365.68	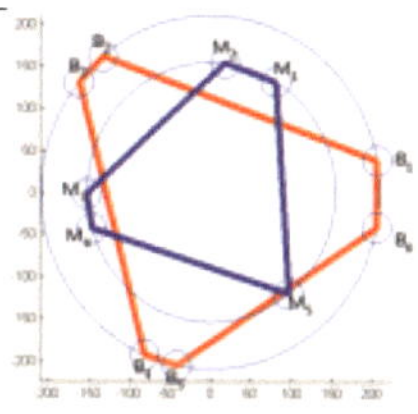
Separation angles of the moving platforms (in degrees)	δ_{m1}	58.48	51.26	101.75	70.04	58.29	Stroke length of 100
	δ_{m2}	83.37	65.95	117.10	76.40	64.93	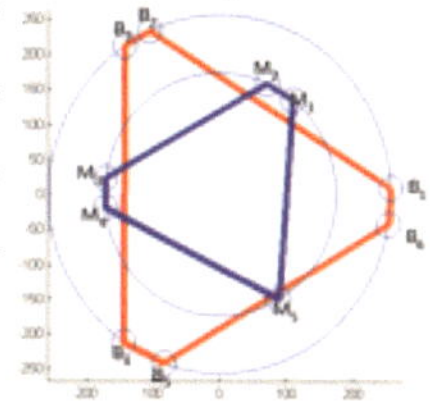
	δ_{m3}	180.54	172.20	131.39	188.63	178.31	
	δ_{m4}	196.26	186.10	231.68	195.02	184.91	
	δ_{m5}	308.59	300.46	345.71	308.83	300.66	
Separation angles of the moving platform (in degrees)	δ_{b1}	10.56	2.32	1.24	22.79	3.56	Stroke length of 150
	δ_{b2}	129.96	114.27	119.10	123.07	119.67	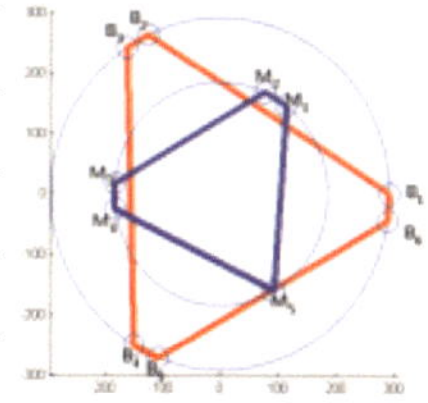
	δ_{b3}	141.44	123.93	232.00	141.77	123.65	
	δ_{b4}	247.00	235.64	239.19	242.34	239.74	
	δ_{b5}	258.49	250.94	256.19	262.16	245.06	
	δ_{b6}	346.24	350.59	351.80	355.45	356.21	
Retracted Length (in mm)		150	250	280	592	740	Stroke length of 200
Extended Length (in mm)		200	350	430	792	990	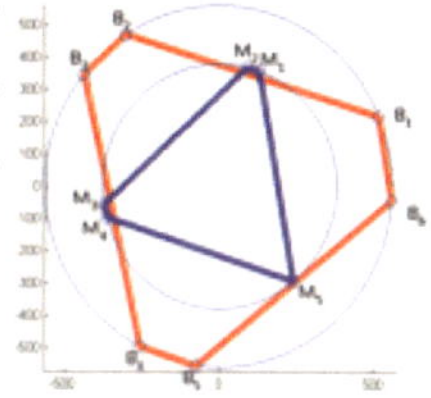
GDI		0.928	0.922	0.915	0.884	0.886	
WSP (cm^3)		106.9	741.2	1531.6	1620.0	1805.1	

Connection points of the legs on base and moving platforms							Stroke length of 250
Legs	L_1	L_2	L_3	L_4	L_5	L_6	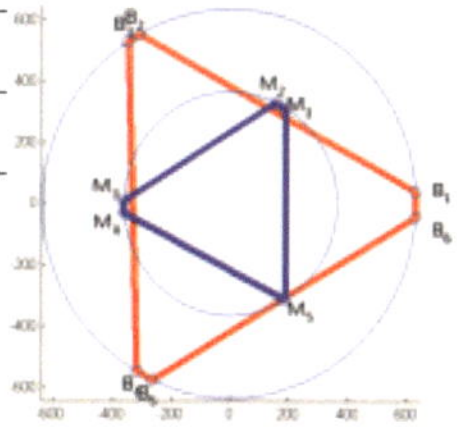
Base	B_1	B_2	B_3	B_4	B_5	B_6	
Moving	M_1	M_2	M_3	M_4	M_5	M_6	

Table 11.
The optimization results for 6x5 GSP.

Optimized design variables		Stroke lengths					
		50	100	150	200	250	
Radius	r_b	205.81	290.40	320.49	599.12	688.71	Stroke length of 50 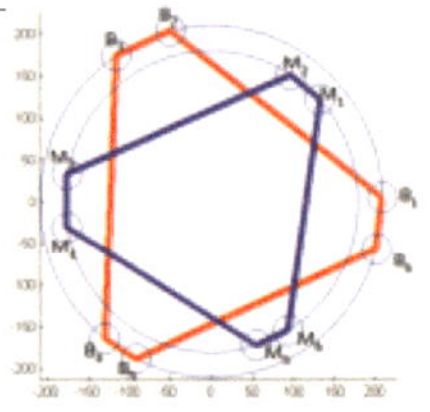
	r_m	153.73	225.11	226.68	434.32	443.86	
Separation angles of the moving platform (in degrees)	δ_{m1}	0.00	13.00	56.05	68.94	11.88	Stroke length of 100 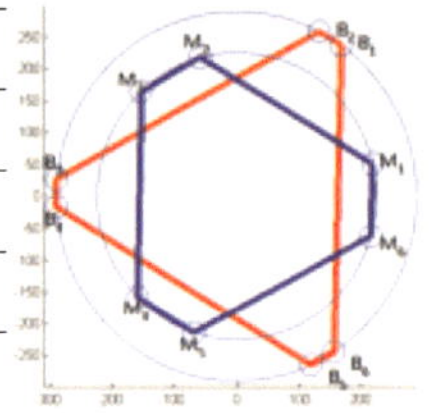
	δ_{m2}	85.69	104.95	68.70	75.73	115.77	
	δ_{m3}	104.87	133.14	175.54	186.96	130.05	
	δ_{m4}	209.34	224.86	187.82	194.65	235.77	
	δ_{m5}	225.38	252.24	296.01	306.33	251.01	
	δ_{m6}	338.46	343.85	307.00	312.08	354.55	
Separation angles of the base platform (in degrees)	δ_{b1}	35.13	54.63	10.02	24.72	62.18	Stroke length of 150 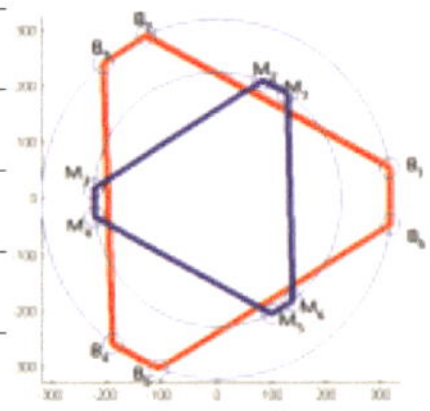
	δ_{b2}	46.84	62.92	113.89	119.63	65.82	
	δ_{b3}	156.76	174.66	130.56	143.06	180.36	
	δ_{b4}	168.89	183.27	233.35	239.03	185.53	
	δ_{b5}	269.71	294.13	250.63	262.41	300.64	
	δ_{b6}	284.75	302.43	351.92	355.88	304.74	
Retracted Length (in mm)		150	250	280	592	740	Stroke length of 200 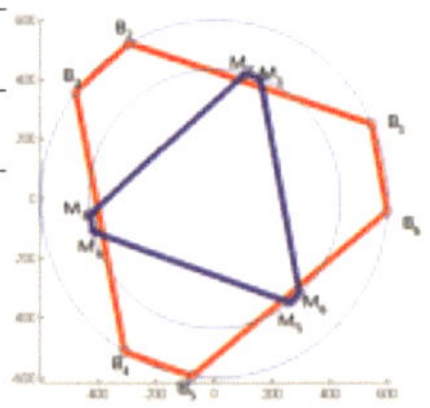
Extended Length (in mm)		200	350	430	792	990	
GDI		0.918	0.923	0.918	0.891	0.893	
WSP (cm3)		116.47	777	1558	1552	1738	

Connection points of the legs on base and moving platforms						
Legs	L_1	L_2	L_3	L_4	L_5	L_6
Base	B_1	B_2	B_3	B_4	B_5	B_6
Moving	M_1	M_2	M_3	M_4	M_5	M_6

Stroke length of 250

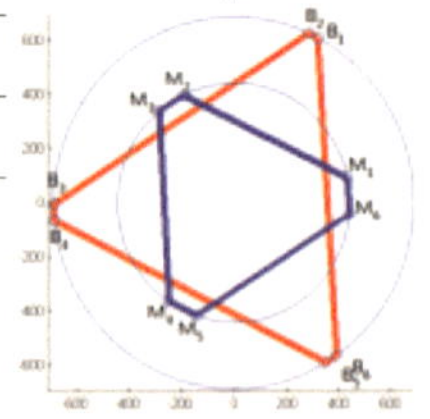

Table 12.
The optimization results for 6x6 GSP.

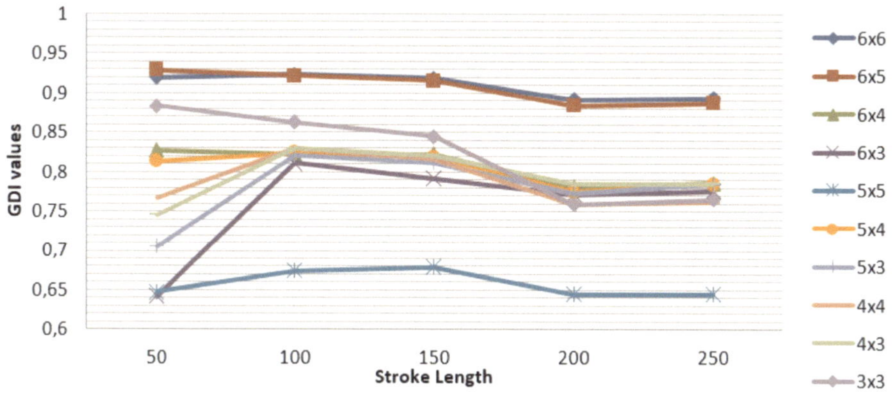

Figure 7.
The GDI values of the ten different GSPs for each actuator stroke length.

actuator stroke lengths in **Tables 14**. The GDI values of 5x5 GSP mechanisms for each actuator stroke length vary between 0,645 and 0,678 which is the worst kinematic performance among the others. In addition, these manipulators have also smaller workspaces that might not be preferred by the designer for constructing their GSPs.

7. Conclusions

Ten different types of 6-DOF GSPs with five different actuator stroke lengths are optimized in this chapter. Dexterities and workspaces of the manipulators are considered as optimization objectives in order to obtain the radius of the circumferential circles, and separation angles between adjacent vertices of base and moving platforms. The PSO is used as the optimization algorithm. The optimization results of ten different GSPs with five different linear actuator stroke lengths are illustrated as tables that include the radius of the circumferential circle, separation angles between adjacent vertices, workspace volumes, GDI values, shapes of the base and moving platforms, retracted and extracted lengths of the linear actuators, and connection points of the legs on the base and moving platforms in order to show the data belonging to the manipulators in a compact form.

The optimizations produced the following important results. The radii of the base platforms are obtained larger than the radii of the moving platforms for GSPs with each actuator stroke length. The locations of the connection points on base platform form a rough triangle. The rough equilateral triangle structure for the base and moving platforms can produce the best kinematic performance like 6x5 and 6x6 GSP mechanisms. These manipulators have the best dexterous maneuverability and kinematic performance, and also the largest workspaces for each actuator stroke length. The actuator stroke lengths of 100mm and 150mm produce the best GDI values for ten different GSPs. The 5x5 GSP mechanisms for each actuator stroke length have the worst kinematic performance and smaller workspaces.

In practice, the designers and researchers can use the optimization results given in **Tables 6–14** to construct the optimal GSP mechanisms for the given specific tasks.

	Stroke length 50				Stroke length 100				Stroke length 150		
Order	GSP	GDI	Workspace (in cm^3)	Order	GSP	GDI	Workspace (in cm^3)	Order	GSP	GDI	Workspace (in cm^3)
1	6x5	0.928	106.88	1	6x6	0.923	777.34	1	6x6	0.915	1558.02
2	6x6	0.919	116.47	2	6x5	0.921	741.16	2	6x5	0.914	1532.00
3	3x3	0.883	75.58	3	3x3	0.862	511.59	3	3x3	0.844	1346
4	6x4	0.827	87.23	4	4x4	0.830	131.88	4	6x4	0.821	1415
5	5x4	0.813	96.91	5	4x3	0.829	328.23	5	4x3	0.820	1133
6	4x4	0.767	48.26	6	5x4	0.823	669.90	6	5x4	0.818	1426
7	4x3	0.745	55.66	7	6x4	0.821	651.22	7	4x4	0.815	747
8	5x3	0.705	58.2	8	5x3	0.820	290.20	8	5x3	0.811	1095
9	5x5	0.648	30.69	9	6x3	0.810	250.09	9	6x3	0.791	1099
10	6x3	0.642	56.88	10	5x5	0.673	644.96	10	5x5	0.678	653

Table 13.
GSPs with stroke lengths of 50, 100, and 150.

	Stroke length 50				Stroke length 100				Stroke length 150		
Order	GSP	GDI	Workspace (in cm^3)	Order	GSP	GDI	Workspace (in cm^3)	Order	GSP	GDI	Workspace (in cm^3)
1	6x5	0.928	106.88	1	6x6	0.923	777.34	1	6x6	0.915	1558.02
2	6x6	0.919	116.47	2	6x5	0.921	741.16	2	6x5	0.914	1532.00
3	3x3	0.883	75.58	3	3x3	0.862	511.59	3	3x3	0.844	1346
4	6x4	0.827	87.23	4	4x4	0.830	131.88	4	6x4	0.821	1415
5	5x4	0.813	96.91	5	4x3	0.829	328.23	5	4x3	0.820	1133
6	4x4	0.767	48.26	6	5x4	0.823	669.90	6	5x4	0.818	1426
7	4x3	0.745	55.66	7	6x4	0.821	651.22	7	4x4	0.815	747
8	5x3	0.705	58.2	8	5x3	0.820	290.20	3	5x3	0.811	1095
9	5x5	0.648	30.69	9	6x3	0.810	250.09	9	6x3	0.791	1099
10	6x3	0.642	56.88	10	5x5	0.673	644.96	10	5x5	0.678	653

Table 14.
GSPs with stroke length of 200, 250.

Author details

Burak Inner[1] and Serdar Küçük[2*]

1 Engineering Faculty, Department of Computer Engineering, Kocaeli University, Kocaeli, Turkey

2 Technology Faculty, Department of Biomedical Engineering, Kocaeli University, Kocaeli, Turkey

*Address all correspondence to: skucuk@kocaeli.edu.tr

References

[1] Tsai LW. Robot Analysis: The Mechanics of Serial and Parallel Manipulators. New York: John Wiley & Sons; 1999

[2] Merlet JP. Parallel Robots. Netherlands: Kluwer Academic Publishers; 2000

[3] Kang BH, Wen JT, Dagalakis NG, Gorman JJ. Analysis and design of parallel mechanisms with flexure joints. In: Proc. of IEEE International Conference on Robotics and Automation. New Orleans, LA, USA. 2004. pp. 4097-4102

[4] Uchiyama M. Structures and characteristics of parallel manipulators. Advanced Robotics. 1994;**8**(6):545-557

[5] Gao F, Li W, Zhao X, Jin Z, Zhao H. New kinematic structures for 2-, 3-, 4-, and 5-DOF parallel manipulator designs. Mechanism and Machine Theory. 2002; **37**(11):1395-1411

[6] Gallardo-Alvarado J, Orozco-Mendoza H, Rico-Martínez JM. A novel five-degrees-of-freedom decoupled robot. Robotica. 2010;**28**(06):909-917

[7] Li F, Kuiper JH, Khan SA, Hutchinson C, Evans CE. OC30 a new method to measure the inter-fracture site movement (IFMS) dynamically by means of Stewart platform manipulator algorithm. Journal of Bone and Joint Surgery - British Volume. 2008;**90-B** (SUPP II):366-367

[8] Pittens and Podhorodeski. A family of Stewart platforms with optimal dexterity. Journal of Robotic Systems. 1993;**10**(4):463-479

[9] Stoughton RS, Arai T. A modified Stewart platform manipulator with improved dexterity. IEEE Transactions on Robotics and Automation. 1993;**9**(2): 166-173

[10] Du Plessis LJ, Snyman JA. A numerical method for the determination of dexterous workspaces of Gough–Stewart platforms. International Journal for Numerical Methods in Engineering. 2001;**52**:345-369

[11] Su YX, Duan BY, Zheng CH. Genetic design of kinematically optimal fine tuning Stewart platform for large spherical radio telescope. Mechatronics. 2001;**11**:821-835

[12] Yao R, Zhu WB, Yang QG. Dimension optimization design of the stewart platform in FAST. Advanced Materials Research. 2011;**308–310**:2110-2113

[13] Mishra A, Omkar SN. Singularity analysis and comparative study of six degree of freedom Stewart platform as a robotic arm by heuristic algorithms and simulated annealing. International Journal of Engineering Science and Technology. 2011;**3**(1)

[14] Jiang Q. Singularity-Free Workspace Analysis and Geometric Optimization of Parallel Mechanisms. Québec: Université Laval; 2008

[15] Jiang Q, Gosselin CM. Effects of orientation angles on the singularity-free workspace and orientation optimization of the Gough–Stewart platform. Journal of Mechanisms and Robotics. 2010;**2**: 011001-1

[16] Jiang Q, Gosselin CM. Geometric optimization of the MSSM Gough–Stewart platform. Journal of Mechanisms and Robotics. 2009;**1**:031006-1

[17] Jiang Q, Gosselin CM. Maximal singularity-free total orientation

workspace of the Gough–Stewart platform. Journal of Mechanisms and Robotics. 2009;**1**:034501-1

[18] Guru S, Halgamuge S, Fernado S. Particle swarm optimisers for cluster formation in wireless sensor networks. In: Proceedings of the 2005 International Conference on Intelligent Sensors, Sensor Networks and Information Processing Conference. Melbourne, VIC, Australia. 2005. pp. 319-324

[19] Rahmat-Sammi Y, Jin N, Xu S. Particle swarm optimization (pso) in electro-magnetics: Let the bees design your antennas. In: 22nd Annual Review on Progress in Applied Electromagnetics. 2006

[20] Selvan S, Xavier C, Karssemeijer N, Sequeira J, Cherian R, Dhala B. Parameter estimation in stochastic mammogram model by heuristic optimization techniques. IEEE Transactions on Information Technology in Biomedicine. 2006;**10**(4):685-695

[21] Shi X, Yeo KS, Ma J-G, Do MA, Li E. Scalable model of on-wafer interconnects for high-speed CMOS ICs. IEEE Transactions on Advanced Packaging. 2006;**29**(4):770-776

[22] Kucuk S. Energy minimization for 3-RRR fully planar parallel manipulator using particle swarm optimization. Mechanism and Machine Theory. 2013;**62**:129-149

[23] Toz M, Kucuk S. Dexterous workspace optimization of an asymmetric six-degree of freedom Stewart–Gough platform type manipulator. Robotics and Autonomous Systems. 2013;**61**(12):1516-1528

[24] Kiranyaz S, Pulkkinen J, Gabbouj M. Multi-dimensional particle swarm optimization in dynamic environments. Expert Systems with Applications. 2011; **38**:2212-2223

[25] Angeline PJ. Tracking extrema in dynamic environments. In: Proceedings of the 6th Conference on Evolutionary Programming. Heidelberg: Springer-Verlag; 1997. pp. 335-345

[26] Kennedy J, Eberhart RC, Shi Y. Swarm Intelligence. New York: Morgan Kaufmann; 2001

[27] Aliyari MS, Teshnehlab M, Sedigh AK. A novel training algorithm inANFIS structure. In: Proceedings of American Control Conference. Minneapolis, Minnesota; 2006

[28] Blackwell TM, Branke J. Multi-swarms, exclusion, and anticonvergence in dynamic environments. IEEE Transactions on Evolutionary Computation. 2004;**10/4**:51-58

[29] Kucuk S. A dexterity comparison for 3-DOF planar parallel manipulators with two kinematic chains using genetic algorithms. Mechatronics. 2009;**19**(6):868-877

[30] Klein CA, Blaho BE. Dexterity measures for the design and control of kinematically redundant manipulators. The International Journal of Robotics Research. 1987;**6**(2):72-83

[31] Gosselin C, Angeles J. A global performance index for the kinematic optimization of robotic manipulators. Journal of Mechanisms and Robotics. 1991;**113**:220-226

[32] Zanganeh EK, Angeles J. Kinematic isotropy and the optimum design of parallel manipulators. The International Journal of Robotics Research. 1997;**16**(2): 185-197

[33] Kennedy J, Eberhart R. Particle swarm optimization. In: Proceedings of the International Conference on Neural Networks. NJ: Piscataway; 1995